The Springs of Steamboat

The Springs of Steamboat

Healing Waters, Mysterious Caves *and* Sparkling Soda

Dagny McKinley

The History PRESS

Published by The History Press
Charleston, SC 29403
www.historypress.net

First published 2013

Manufactured in the United States

ISBN 978.1.60949.880.1

Library of Congress CIP data applied for.

This book is dedicated to the mineral springs of Steamboat Springs—for what they have taught me about the history of this town and for their patience with the decisions of mankind.

Contents

Acknowledgements

Thank you to Edith Hornik Beer, without whose encouragement I would never have acknowledged myself as a writer. A huge thank-you to the Steamboat community for their help and story contributions. Thanks to Bill Fetcher for providing historic photographs of the Strawberry Park Hot Springs. To George Tolles for his personal notes, time and help. To Jim Crawford for helping me make the manuscript as accurate as possible with the knowledge he had, for his time and his willingness to review my manuscript. To the Tread of Pioneers Museum for all its help and patience as I researched this project. To Joe Stepan at the Strawberry Park Hot Springs for providing so much information. To Pat Carney and Jeanne Gillaspie for their contribution of historic photographs and information on the Old Town Hot Springs and Steamboat Spring Health and Recreation Association. To the *Steamboat Pilot* for the mass amount of information, without which this book would not have been possible. To Tom Wither for sharing personal family stories and photographs. To Corey Kopishke for the beautiful summer scenic shot of the Old Town Hot Springs. To Jody Condie for sending a copy of the postcard of Steamboat. To Suzie Hawkins, Rodney Pryde, Cookie Lockhart, Chris Erickson and Rick and Linda Banter for their story contributions. To Emily Katzman, without whose help this book would never have been finished in time. To Kim Keith for her contribution of photography and friendship. And to my family and my dog for their patience as I worried my way through every word of this project.

Introduction

The town of Steamboat Springs was named for a once-famous spring that sounded like the chugging of a steamboat. Steamboat was built close to the mineral springs for all the wonders they offered, from drinking water to bathing opportunities to medical remedies. Today, when people hear the name Steamboat Springs, they think of a world-class ski resort. They think of hiking, biking, fly-fishing and access to endless outdoor recreational opportunities. Very few people today come to visit Steamboat for the healing powers of the mineral springs. Many may spend an afternoon or evening at the Old Town Hot Springs or the Strawberry Park Hot Springs, soaking in the waters without realizing the history of the springs. These two springs have been developed as focal points for recreational activities while the remainder of the springs sit, unassuming, at the edge of town where they often go unnoticed.

Just a little while ago, a visitor to Steamboat asked me how the town got its name. Most people who live here know the story of the Steamboat Spring, but until I began this book, I had never visited that spring or the sulphur spring that offers a very distinctive greeting to those driving into town. I had no idea that the springs here had so much to offer. As the town grew up, people like me, without any emotional ties to the springs, made development decisions they thought best for the direction of the town. In the process, some springs disappeared.

I have lived in the Yampa Valley for over ten years. Ten years before that, I was an occasional visitor to the area. I know all the runs on the ski mountain.

Postcard showing Steamboat and Old Town Hot Springs. *Courtesy Jody Condie.*

I have both hiked and skied them. I know what the snow conditions are like daily but didn't know that Steamboat was once a destination town for healing, with people traveling across the country solely to visit the 150 mineral springs found here. The first encounter with the valley was often a soak in the Old Town Hot Springs, or Bath Spring, as it was once known. Visitors came to drink from the Lithia Spring as a cure for all sorts of physical ailments. As the town developed, ranchers moved in and became the primary industry in town. Carl Howelsen introduced skiing for recreation and ski jumping in 1914 when he built Steamboat's first ski jump. With that, he changed the course of this town's history. The focus for tourism became skiing. Every winter, people flock from around the world to try Steamboat's famous champagne powder™.

Recently, traditional weather patterns have been changing. The seasons appear to be shifting, with snow falling later in the season. Weather cycles of drought or extreme moisture are not uncommon. In dry years, the tourism industry suffers, not to mention the ranchers and townspeople. So Steamboat is yet again adapting by building a reputation as a biking destination in the summer.

With each new fad, older ones fade into the background. The activities that draw in the biggest crowds tend to get the most dollars in regard to expansion, preservation and attention. There have been periods in Steamboat's history when some of the mineral springs were allowed to fall into disrepair. Fortunately, there have always been some strong voices in the community that have fought for the preservation and protection of our springs. One of those voices was that of Lulita Crawford Pritchett, granddaughter of the founding mother and father of the town, James and Margaret Crawford. Another crucial person who has given a voice to the springs is George Tolles, who drinks daily from the Lithia Spring and soaks regularly in the Old Town Hot Springs. Dorothy Wither was a proponent of the Lithia water and the Old Town Hot Springs until her passing.

Without the mineral springs, Steamboat would never have become the town it is today. To visit the various springs now is at once to see the history of this town and the future. In the 1950s, children, including Tom Wither, set up a volunteer-guided tour of the various springs. Their headquarters were at the Soda Spring. They had a big sign that read "Free tours of the hot springs." The boys researched about five of them and read the information to customers. "It wasn't a big thing," remembers Tom, "but every now and then tourists would call or come by and give us a tip of a dollar or something." Currently, the Tread of Pioneers Museum downtown offers a free walking tour of the mineral springs (excluding Strawberry Park Hot Springs) once a week during the summer months.

My hope is that this book will stand as another voice for each of these springs. These words hopefully tell an accurate history of these waters while leaving open the possibility that we may still not know the extent of the beneficial attributes of these springs, how they serve the environment and how they can heal—mentally, physically and spiritually—those of us fortunate enough to call Steamboat home.

Chapter 1

A Brief History of the Mineral Waters of Steamboat Springs

Overview

In all ages humanity has placed a high value on mineral springs. In the long twilight ages, not knowing how else to explain their soothing and healing powers, they were said to be peopled with spirits from the under world, of which origin their heat, bubbles, gases, and sulphurous smells were, to the savage mind, clear and convincing proofs.

—Steamboat Pilot, *Recreation Edition, 1940*

The main interests of the people who populated and continue to populate Steamboat Springs and the surrounding valley directly affect the way the various springs are treated. Despite the mineral waters' place in history as one of the key reasons this valley was coveted and then settled, certain springs have been highly regarded while others have, at times, been ignored. Some have been buried or lost to development or cattle stampedes. As progress, expansion and development shaped Steamboat Springs, ideas about the importance of the mineral springs have also changed.

Humans have a tendency to label people, animals and landscapes as important or not based on what they can provide physically, aesthetically and financially. Because each generation understands more than the generation before it, we may not for some time fully grasp the consequences of man's interference with the various springs in this town. In the past, "progress," or development and exploitation of resources, was put ahead of just about anything, including preservation of the landscape surrounding the mineral springs. However, there have always been visionaries who have spoken up for the preservation of this town's natural resources. Because of them there are springs that are still revered and cared for.

Over time, this valley has transformed from a land where migrating American Indians summered each year to a trapper's paradise to settlers, ranchers and miners staking claims and finally, now, a haven for skiers, boarders and lovers of the outdoors. With each influx of interests, the town has been shaped and marks left, heritages established, history gained or lost. By following the story of the mineral springs, their place and man's place in this valley becomes marked.

PHYSICALITY

Long before people inhabited Steamboat Springs, waters flowed under and above ground. Over 150 springs sprouted up into the landscape and flowed back into the Yampa River. Each mineral spring varies in size, shape, flow, smell and taste. Some springs offer up a tiny trickle while others spout forth hundreds of gallons of water per minute. Temperatures also fluctuate with each spring. The lowest temperature recorded in any of the springs is a cool 40 degrees, while the steamiest is heated by geothermal heating to approximately 150 degrees. Colors are as wide-ranging as temperatures and reflect the mineral content of each spring. From black to white and almost every shade of blue in between, each spring has a unique personality and a unique purpose. Mineral content also determines the smell and taste of each spring. The most commonly found minerals include Lithia, iron, calcium and sulphur. Looking at them, some are more agitated than others. Some springs gurgle, while others bubble. Some soothe, others excite. Their minerals offer nutrients to wildlife and humans savvy about their healing powers.

Thermal springs, like the Old Town Hot Springs and Strawberry Park Hot Springs, are believed to sprout up from various fault lines surrounding the vicinity of Steamboat. A once-popular belief claimed that no two Steamboat springs were exactly alike. However, this theory was proven untrue after a study prepared by the Colorado Geological Survey in cooperation with the U.S. Department of Energy. The report concluded that the Heart Spring at the Old Town Hot Springs is hydrologically related to the Routt Hot Spring (or, as it is now known, the Strawberry Park Hot Springs). Thermal waters found during the construction of the high school in Strawberry Park appeared to further confirm this by showing that the waters from Strawberry Park Hot Springs flow down toward the Old Town Hot Springs. Geological mapping shows a major fault that runs from the Routt Hot Springs Park. The thermal waters could potentially serve a purpose other than recreation.

The Colorado Geological Survey, undertaken in 1983, was conducted to evaluate alternative resources for energy. The study believed geothermal energy would at some time in the future become "an important source of local energy for…heating and cooling of buildings, processing food, heating cattle barns, greenhouses and fish ponds, milk pasteurization; and recreation and medicinal purposes." The report found:

> *Precambrian age rocks have been thrust over Jurassic and younger rocks along a thrust fault north of Steamboat Springs. Tweto (1976) shows the continuous north–south normal fault cutting the area to be part of a fault system that extends for almost 50 miles along the west side of the Park and Gore Ranges. In the Steamboat Springs area, the Browns Park Formation has been downdropped along this fault into contact with the Morrison Formation. The major north–south fault consists of several smaller parallel faults. Geological mapping by Snyder (1977 and 1980) showed that the thermal springs on the north side of Steamboat Springs are the only ones directly associated with any faults.*
>
> *From the geological evidence available, it appears that the Routt Hot Springs are fracture controlled and showed a north–south trending fault passing just to the west of the Routt Hot Springs. Snyder (1980) showed that in the vicinity of the Routt Hot Springs the strike and dip of the metamorphic schistosity or igneous foliation is highly variable suggesting that the bedrock is highly fractured. Numerous fractures abound in the vicinity of the thermal springs.*

Only three of the springs within the township of Steamboat are considered thermal: the Old Town Hot Springs, the Sulphur Cave Spring and the Steamboat Spring. The Strawberry Park Hot Springs are the hottest of all the springs but are not included within the township of Steamboat. The Heart Spring was found to be fault controlled. As for the Steamboat Spring and Sulphur Cave Spring, their waters may come from faults that cut through the Dakota sandstone.

Most thermal waters are meteoric in origin, meaning they have come from precipitation above ground and then have been absorbed into the groundwater system. While most of the precipitation in Steamboat ends up in the Yampa River and other streams of water, a small portion ends up in the groundwater system that "circulates downward to depth along the many faults and fractures in the area where it becomes heated," according to the Colorado Geological Survey.

The Strawberry Park Hot Springs, located outside town, also qualify as thermal springs. In 1983, the springs still belonged to the city and remained undeveloped. The Strawberry Park Hot Springs were measured as having a temperature between 124 degrees and 147 degrees Fahrenheit. The area of Strawberry Park, where strawberries were once grown, has a higher ground temperature than the rest of Steamboat, leading engineers to believe that water runs from north to south. In winter, snow melts more quickly and roads clear and dry faster in Strawberry Park than in other areas of town.

Local geologist, scientist, rock climber, hiker, skier, Harley-rider, water skier and photographer Bill Bowes believed the springs got their heat from Hahn's Peak, a young volcano located approximately thirty miles north of town. He believed (as quoted in the *Steamboat Pilot*) that the water was heated there, flowed through fault lines and then came back to the surface. This was a popularly held belief for much of Steamboat's history. However, later reports claim that a possible source of heat is the disintegration of radioactive minerals. Front Range rocks of the Tertiary age in the Colorado Mineral Belt are up to twenty-five times more radioactive than other granitic rocks. Northeast of Steamboat, the presence of radioactive minerals has also been found. "Therefore it can be assumed that some heat could be contributed by decay of radioactive minerals in basement rocks," reported the Colorado Geological Survey. The report also concluded that in order for the Old Town Hot Springs and Strawberry Park Hot Springs to reach the temperatures they do, they most likely dip down to twelve thousand feet below ground. The water content of the two springs was shown to be almost identical through water analysis. The Old Town Hot Springs are the only springs that do not appear to lie on a fault.

The fault lines are believed to be responsible for the plethora of springs in this area. If an earthquake or a shifting of the earth occurs, the possibility is good that some springs might disappear back into the earth.

Those visiting the springs today will note not only the variance in temperature but also a great variance in taste, appearance and physical makeup. According to George Tolles, an advocate for the mineral waters of Steamboat, "What springs are called depends on the kind of rock environment they are in. They take the characteristic of the rock they flow through." People and animals also adapt to the characteristics of landscape and the natural resources those areas provide.

Inhabitants

Before Steamboat had human inhabitants, trees grew high and wild and all variety of wildlife wandered through what is now known as the Yampa Valley. Grizzly bears, wolves, black bears, foxes, mountain lions, elk, deer, grouse and big horn sheep all roamed through these hills taking what they needed to survive, marking territories and feasting on the salty grasses surrounding the mineral springs. They quenched their thirst in the sulphur springs and the Yampa River. Wildlife was drawn to the two springs of the Kissingen-saline type. Even horses and cattle were drawn to these waters. In later years, heavy grazing and trampling by livestock caused a few springs to disappear.

Grizzly bears also frequented the springs to get nutrients from the various minerals. Before people settled the area, grizzly bears made up a large portion of the population in Steamboat. There's a story of Logan Crawford, one of the children of the first family to permanently settle the valley, running into five grizzlies in a single day. He hid himself until the bears moved on, but the encounter was enough to put him on edge.

The Ute Indians are said to be the first people to spend time in the valley. Scientists say the Utes migrated through the Bering Strait into the Rocky Mountain area. The Ute Indians have a different explanation for their presence in Colorado: the Rocky Mountains have always been their home. This is where the Creator placed them and where they have always existed, according to northern Ute historian Roland McCook. For a time until the white man came, the Yampa Valley remained the summering place of the Utes.

The Utes were dark-skinned American Indians. Their skin was so dark that they were referred to as "black Indians" by other tribes, according to Wilson Rockwell in his book *The Utes: A Forgotten People*. The northwestern Colorado tribes, according to Rockwell, were divided into three bands: the Grand River, Yampa and Uintah.

Approximately five to six hundred Utes would arrive in Steamboat for the summer season. After the long journey walking and riding horses from southern Colorado, Ute elders would soak in the hot springs and sulphur springs to heal their bodies. The first whites to arrive, most likely trappers or explorers at the time, would be invited to share pipes with the Utes to prove they were friendly. The lush valley offered hunting opportunities; space for horse racing, of which the Utes were fond; and access to "healing waters," as they referred to the springs.

Magic was in the healing waters, the Utes believed. "Steam coming out of undeveloped areas of the earth…water is the blood of mother earth and is very, very special," said McCook. According to his oral history recorded at the Tread of Pioneers Museum, Utes "would bathe in the sulphur springs and wear the minerals on their skin. They would bathe horses prior to hunts or battle with other tribes as a way of blessing the horses. The belief was the sulphur gave the horses magical powers to go into battle, like stamina and surefootedness." After battle, the hot springs were visited to rejuvenate strength and spirit.

According to Rockwell's *The Utes*, prior to the introduction of horses, Utes lived and traveled in small family units, with each family hunting and harvesting foods from different areas of the state. By doing so, resources and wildlife remained plentiful for each family. The patterns of where each family traveled were so well established that in times of war, families could easily be found to band together and fight enemies. Once a year, the Utes would gather for the Bear Dance, where romances began, marriages took place and couples decided which side of the family (the bride's or the groom's) they would join and travel with.

When the Spanish introduced horses, the Utes were able to hunt buffalo in groups. Buffalo skins were considered superior to deer and elk for warmth, bedding and teepees. On horseback, the Utes were able to travel farther in less time. Individual family units were replaced with larger tribes. Hunting parties went out on raids of enemy tribes, which included the Arapaho, Cheyenne, Kiowa, Comanche, Sioux and Pawnee. Buffalo hunts sometimes turned into raids on other tribes. Not all surrounding tribes were considered hostile. The Utes maintained friendly relations with the Snake, Shoshoni, Bannock and Paiute.

One of the results of having larger tribes living together was that resources were used up more quickly. Animals in areas were hunted in greater numbers than before. When trappers showed up with goods for trade, such as sugar and whiskey, more animals were killed in order to trade hides.

As trappers moved west, they added to the strain on hunted animal populations. There was a twenty-five-dollar bounty on the head of a cougar. Wolverines, grizzly bears and wolves were hunted to extinction in Colorado after the arrival of white settlers. Coyotes, which had never been seen in Steamboat in early years, migrated to Colorado to replace the wolf population and fill their place on the predatory chain. Fox and mink were also populous. Today, mink are a rare find, but fox, coyote, deer, elk, moose and black bear are plentiful.

With animal pelts bringing in a decent living, animals began to be valued for skins rather than nutrition. Whereas before the Utes used every part of an animal, white people came and took only what they considered valuable. Whites and Utes now began to come into contact on a fairly regular basis. Encounters were not always peaceful.

Americans from the East were heading west in search of gold, silver, precious metals and minerals. Miners didn't care where the minerals were found, even if they were on lands considered sacred by American Indians. Land that had once belonged to the American Indians was now being claimed by strangers, and the Utes were being pushed off lands that were familiar sources of survival for them.

Utes "look at the mountains as providing food, shelter and water. The mountains are sacred. Look at a stand of trees and see shelter for deer or elk," said Roland McCook. Native Americans looked at the way the earth provided for each living thing as sacred. The earth itself was sacred. Europeans looked at the mountains and saw riches within. Their mindset was that God had given the earth to man and that earth was here to be conquered by man. Despite the disparity in views, for quite some time, the Yampa Valley remained a place where peaceable relations were maintained with travelers passing through and the earliest settlers.

Before the white man came, the valley and the main river running through town were called the Yampa River for the Yampa roots that squaws pounded into meal. White men renamed it Bear River for the abundance of grizzly bears that once roamed the area. After the bears disappeared and in honor of the Ute Indians, the original names for landmarks in the town were eventually reinstated.

The white men who traveled through the valley were struck by the lush surroundings and the few springs that had been discovered up to that point. One of the earliest mentions of Steamboat Springs by a white man dates from 1843, thirty-two years before the first permanent white settlers arrived.

"The Diary of Early Adventurer Describes Springs in 1849" was published in the *Steamboat Pilot* on Thursday, June 16, 1988:

Portia Mansfield [, known for instituting the Perry Mansfield Performing Arts Center with Charlotte Perry, discovered and] *gave the* Pilot *a copy of an extract from the diary found in a collection of an old Baltimore family. The friend who made the copy for her said that in the early '49 days a group of young Baltimore bachelors came west looking for gold and adventure. The following paragraphs were taken from the journal of Dr. Brierly of the Charlestown, VA., Mining company.*

"Wednesday, July 11, 1849. We were all aroused this morning in the midst of sweet sleep. By taking a cut three miles brought us upon the road which ran into a beautiful valley skirted on each side by rolling mountains, the verdure extending to the very tops without a rock or shrub to break the velvety surface.

"Twelve miles brought us again to this river and to the well renowned lava springs, by some called Bear springs. We arrived at day-break and corralled. The teams we had been with for several days expressed surprise to find us before them, and snugly and pleasantly laid up in this delightful spot. The whole surface of the earth for miles around shows the effect of one immense or many volcanoes. Along this lava spring valley the earth is covered with charred eruptive stones. In many places the earth is burst up as with an eruption very lately. In other places the rocks and earth are completely split open and you can look as deep down as the eye can penetrate. Down many of them you can distinctly hear running water, showing that a long time ago there was at least a great commotion in these parts. I cannot attempt to account for the many different freaks and irregularities and demands of departed ages that we saw at this place. The whole valley is the most interesting spot of earth I ever beheld. Here is a grand field for the geologist, mineralogist and naturalist. The road here crosses a little creek which empties 200 yards from crossing into bear creek. Immediately at its crossing we found two springs, both of which were soda. They arose at the edge of the water. We found a great number of these along the banks and also in the bed of the creek. You could distinctly see the little springs shooting up. A little farther you find one covering a quarter of an acre. This is called the Boiling spring, it boils up from crevices in the rock in a thousand places, making the surface form a rip. At the lower part of the spring the water [spouts]...*two or three feet high, there is a little intermission of a few seconds every now and then, which makes it resemble the "palaces of the Deep." A few feet from this large one are two smaller ones which are fizzing away all the time and somewhat resemble the escape pipe of a steamer. The large one also has suction power. Someone reached a cup into it, when it*

was immediately drawn from his hand into the sole. He delved down for it and found it the length of his arm in, and it required a considerable jerk to get it out. This soda water is pleasant to drink and when mixed with acid effervesces prettily. It has been analyzed by many, but I do not remember all of its constituent parts. The most important, judging from the deposits around, is carbonate of lime. Fifty yards from the Steamboat spring, just upon the side of the river, were two springs just a foot and a half apart, one of which was a beautiful clear spring, which was very good water and the other perfectly red and copperish.

"There is another spring in the vicinity which our guide told of, the most remarkable one yet. He says that some years since, having lost his cattle, he went out to hunt for them and becoming thirsty he started to look for water. Having found a little trickling in a ravine he followed it up to quite a large spring, which upon approaching, he was surprised to find himself almost suffocated. He stopped a moment, then proceeded a few steps farther, when he thought he would certainly fall if he took another step. He immediately fell back some paces to look for the cause of this remarkable phenomenon. He was astonished to see around the edge of this spring numbers of dead birds, rabbits, frogs, etc. He did not approach nearer, having attempted it several times, and found it would certainly take his breath.

"To the right are the remains of an old crater. So far from having anything like a trace of fire about it, it has every appearance of having been formed by water. It is upon a knoll which covers about an acre. The whole of this knoll being about 50 feet high, of a yellow appearance, resembling at a distance yellow clay, slimy, shiny and greasy and when we walked over has a hollow sound. We took some of our horses over it and in some places the crust broke through, showing a scurf of crust some inches thick. The whole hill was but a shell and when broke through it was a hot, whitish mud, soft and thick. Wherever it was broke, water came out in a stream. Upon this hill was a round hole some six feet in diameter with a wall solid and smooth both inside and outside, coming up above the earth three feet. The inside was filled up to within four feet of the top with dirt, stones and rubbish. Twenty feet from this, on the same mound was a small knob, 4 feet high and 10 feet in circumference, on the top of which, from a hole 6 inches in diameter poured forth a pretty little stream. It bubbled up beautifully and was warm with a sweetish taste. From the top it flowed over this little knob, leaving a deposit of whitish scum, proving to be salaratus. There is no doubt that the whole of this hill has been formed in this way. I examined all of these different places with considerable interest and confess I was fully repaid, and think it would almost repay one

> *the trip across the plains to spend some weeks in this spot. It is a matter of astonishment that no Indian or mountaineers or adventurers have ever settled upon it. It is a beautiful valley, fine water and plenty of timber. The only drawback whatever is the injurious effect of the water upon the stock, which I think with a little care* [could] *be easily prevented. Some of our men were so pleased with it as to determine at some future day, if fortune should fail them, to return to it and be hermits."*

Thirty years later, the call to start a future in the Yampa Valley was shared by Steamboat's first permanent white settler, James Crawford. "Steamboat Springs and Its Founder, an Interview with James H. Crawford" (edited by James Logan Crawford), recorded on May 24, 1923, by Thomas F. Dawson of the Colorado State Historical Society, retells his adventure:

> *His* [James Crawford's] *first visit was made in 1872 and was the result of a newspaper article relating an interview with* [Joseph] *Westcott, who was known to all old timers as Judge Westcott.*
>
> *Judge Westcott was a hermit and for many years was the sole resident on the shores of Grand Lake. He was as familiar with Northwestern Colorado as any man of his time, but owing to the fact that the whole region was claimed by and occupied by the Ute Indians he had not explored it thoroughly. He lived in a little hut on the shores of the lake and although he appeared to be a very commonplace sort of person he seemed inspired by the beauty of the water and the surrounding mountains. He was hospitable to a degree and was unfailing in his efforts to entertain visitors. As going to show his generosity Colonel Crawford relates an incident connected with his first visit to the hermit. It was after a long winter and the old man had exhausted his supplies, and to meet the demands of his hunger he had amputated and cooked the tail of his still living burro. This was all there was between him and starvation and yet he volunteered to share it with visitors.* [J.H.C. wrote in a letter that the visitors included Len Pollard, afterward clerk and recorder of Grand County and a reliable man. Pollard told the story to J.H.C., who did later see the tailless burro.] *His principal claim to fame is the fact that he was the author of a poem which had quite a run for a time with visitors to the lake, to all of whom he read it. It was narrative in form and in it he told the story of a terrific battle on the shores of the lake between the Ute Indians and their enemies of the Plains, the Arapahoes and Cheyennes.*

Westcott read this poem to his visitor of 1872. At any rate, the visitor, who is nameless, wrote an enthusiastic article for the old "Missouri Republican," afterward called the "St. Louis Republic," telling of his visit, expatiating on the beauties of the immediate surroundings and hinting at the wonderful possibilities of the country beyond. He quoted Judge Westcott as saying that the land to the westward was a paradise in appearance and full of undetermined possibilities for the future. This picture attracted Colonel Crawford.

Crawford eventually made his way out to Colorado to find a new home for his family. During the journey, he encountered the buffalo so prized by the Utes, of which he killed plenty:

I remember that on one occasion I drove a big bull buffalo into the camp where the women and children were, and shot him down for their edification. We skinned him and cut him up right there and had some fine pieces of steak cooking shortly afterward. The animal was a splendid specimen. He possessed a fine forelock which we cut off and saved and sent back to friends in Sedalia, telling them that it was an Indian scalp, which they believed and on which account they treasured the trophy for years.

The white man's attitude of domination over the land can be seen in the next lines of Crawford's diary and shows how different the attitudes toward nature were between the Utes and the white settlers:

Passing on, we explored Middle Park, going down the Grand until we were stopped by the Gore range. We only got a look into the region beyond. All that I saw looked good and from that time forward I was determined to go in and "possess" the country…I have said there was no road beyond the springs at Sulphur, but as a matter of fact Sir George Gore, the wealthy and intrepid Irish explorer and big game hunter, had gone into that region with teams and a large retinue of followers some twenty years before on a sporting expedition and he had opened a way through the park and into the mountains, a range of which has since been given his name. [Sir George Gore was an eccentric Irish nobleman who went on a three-year hunting expedition in the 1850s that was more of a killing spree. Some say the slaughter was meant to wipe out the wildlife the American Indians relied on. In a single day, Gore and his party killed two thousand antelope in Middle Park and were responsible for the deaths of over three thousand buffalo.] *There was a trace of his road left, but it was of little use in traveling after twenty years of growth and change…We followed the Gore trail to Egeria Park and there it came to an end. There also I first heard of Steamboat Springs. Before that I was entirely ignorant of the existence of such springs as there are at the place and even then I learned only of one spring, when, as you know, there are dozens of them.*

The information came to me through a wandering prospector, a lone gold hunter who had heard of the "finds" at Breckenridge, Colorado, while following his calling in Utah. Drawn by the new discoveries as is the wont of prospectors, he had found his way through the mountains, carrying his pack on his back, and had chanced to cross the Yampa in the Big Bend, where the springs are located. We met him in the park and he told us of the Bubbling Spring, mentioning it alone. He was a crude sort of fellow, but he had some appreciation of the place for after telling about the spring he said that he would go back there and file on the land if he thought it possible that a wagon road ever would be built into the country. "It is real wonderful" he said; "it smells awful but it's sure fine to look at." After that he went his way and we never heard of him again. If we were told his name I have forgotten it.

Finally, Crawford arrived:

Elk River was up as a result of melting snows, and as we were compelled to wade the stream, which was waist deep, we got well soaked. All told, it was a hard trip, but we got through without accident and, after seeing much country that appealed to us, on the second day out we came to the big spring that the prospector had told us about. It was the spring that has since been called the Bubbling Spring and its appearance was such as to justify the fellow's characterizations of "wonderful" and "smelly." The water is well saturated with sulphur and it continues to the present time to attract the attention of the olfactories. We had been on the ground only a short time before we discovered that there was not one only, but many springs—a very nest of springs. We found probably twenty out of the one hundred and fifty which have since been located, among them the spring which has given name to the place, Steamboat Spring.

After wandering about on the north side of the river on which we had come up, Richardson and I waded the stream to the south side. If I was impressed by my first glimpse of the spot, every successive change or view

> *increased my enthusiasm, and I said to Richardson, "This is the place for me. Let's take up the ground now, come back as soon as we can and make a start." I was satisfied that the springs would insure a future and I saw with my mind's eye a prosperous city at no far distant date. I was sure that this land would be opened and that we would get a title some day.*
>
> *The proposition did not appeal to my companion. He said that he had had about enough of the mountains and had determined to return to Missouri. He added, "Now that we are here I will help you in what you want to do, but Missouri is good enough for me." I accepted his proffer of assistance and we went to work to build the foundation for a house, the land hunter's first sign of a claim. There was a pile of slide-rock near and we pulled enough of it up to the chosen spot to build a wall three or four feet high about an enclosure fourteen by eighteen feet. I wrote an elaborate claim notice and placed it in an inside corner of the wall and in addition peeled the bark off a quaking aspen and wrote a notice on the tree for one hundred and sixty acres of land with the tree as a center of the tract.*

According to Charles H. Leckenby's account of "The Founding of Steamboat Springs and Hahn's Peak" from his book *The Tread of Pioneers*:

> *It was the first day of May, 1873, that Mr. and Mrs. Crawford, with their three small children, left their home in Sedalia, Missouri, to start the search for a new home in the west. They made the trip across the plains by ox team with several other families and arrived in Denver on June 4th. Indians stampeded their horses on one occasion and the high streams and quicksand made the journey to the new land one to be remembered. The Crawfords camped on Cherry Creek but remained in the vicinity of Denver but a short time. They wanted to get into the mountains and continued on to Empire, where they remained until December and then went back to the Culver ranch, three miles above Beaver Brook, and remained until June 8th of the next year.*
>
> *Lured by the unknown country that lay across the continental divide they set forth for Middle Park. On the tenth day of June the Crawford party camped at Yankee Doodle Lake. John Q. Rollins was making a road across the mountains where the Moffat road crossed the divide before the Moffat tunnel was built. The grade on the road was very steep. The Crawfords had a pair of mules and a span of horses and Mr. Rollins hitched his yoke of oxen on to pull the wagon over the mountains. Because of the steepness of the road Mrs. Crawford and her three children, one seven years old, one*

four and one a year old, were forced to get off the wagon. A sudden snow storm came up while the ascent was being made and the children became frightened and started crying. Mrs. Crawford found a saddle blanket and all covered their heads and waited until the storm had subsided. No heavily loaded wagon had made the trip over the mountain range by that route before Mr. Crawford made the perilous trip. He had with him eight cows and calves in addition to his wagon heavily loaded with equipment.

Finally the party landed at Hot Sulphur Springs, where the Crawfords camped until a cabin was built. Mrs. Crawford relates that the cabin roof was covered with bark until Mr. Crawford could find a good tree to rive some boards to cover it. During the summer, she says, for about a month it rained almost every afternoon and the inside of the cabin was about as wet as the outside. The children developed scarlet fever and a wagon sheet was hung over the bed to keep the children dry. A carpet for the dirt floor was made of elk skins with the hair all turned one way so it could be swept.

When James Crawford returned to Hot Sulphur Spring, he shared his vision of Steamboat with his wife, Margaret:

The manifold charms and prospective wealth of the Yampa River valley convinced Mr. Crawford that it would not be possible to find another place possessing so many features of value to the home seeker, and on returning to Hot Sulphur Springs, he announced his intention of conducting his family to the wonderful springs in the Yampa Valley and taking permanent possession of the land he had located there. It was too late in the season to carry out this plan before winter might be expected to set in, as there were no roads leading into the country, which was then an almost unexplored and entirely unsettled region. Mr. Crawford spent the winter about twenty miles down the Grand River where he put up some hay for his few cattle.

Long before the intended journey could be undertaken with safety, Mr. Crawford received information that his location in the Yampa Valley was in danger of being jumped. In order to frustrate the intended jumpers, Mr. Crawford and Lute Carlton made a hurried trip to the new home to perfect Mr. Crawford's right to the land. Although it was in June, 1875, the two men had to buck their way through the snowdrifts on the Gore range and they plodded along day after day in the face of great danger to make the trip. When they arrived at the Yampa River springs Mr. Crawford found no sign of the jumpers. To strengthen his title he added something to the cabin begun the year before, and left other evidences of occupation, including a "garden," which was made on a small spot of bare ground with seeds brought along for that purpose. In order that there would be no misunderstanding concerning the purpose of this bit of cultivated ground it was decorated with small cards, on which was written: "This is lettuce," "these are radishes," etc. These improvements made, the two men started the return trip to Hot Sulphur Springs and again had a trying journey because of the high water.

The story continues with Crawford's interview:

That my trip was not in vain was afterward proved for the Georgetown men told me later that they had had an eye on the place and had fully decided to go after it if I had not followed up my discovery.

The journey was not without its hardships. I will relate one incident which, although it may appear insignificant now, is still an illustration of the vicissitudes to which the pioneers in the Rocky Mountains were subjected. On the way in we made some flapjacks at one of our camps but burned a number of them so badly that we threw them out in the snow. Before our arrival at this place on our return, Carlton and I had exhausted our supply of flour. We were hungry as well as tired and were without bread. We remembered the burned flapjacks and I said that I believed that I could find them. And after a little hunting around under the newly fallen snow I succeeded, and I can tell you that I believe I never tasted more delicious flapjacks in all my life than were those burned specimens. They were fine.

At the time Crawford staked his claim, there was one other sign of settlement. An old cabin had been built near the Sulphur Cave Spring but was already caved in. A yoke from a team of oxen also remained but had grown into the nearby tree. It is believed the man who built the cabin was from Wisconsin and had gotten lost on his way to find the California gold fields in 1849. He built the cabin to spend the winter in and wrote to his son of the very "curious springs" in the valley he was inhabiting.

The idea of white ownership did not sit well with the Ute Indians. This had been their land since the time of creation. Chief Yarmonite, the head chief of the Yampa Valley Utes (sometimes spelled Yahmonite or Yarmony), shared the view of some of the other Ute elders, including the famous Chief Ouray, that the Utes should attempt to live in harmony with the white people. Other younger braves, having heard stories of the Utes being removed

Ute Chief Yarmonite. Sometimes described as a Ute sub-chief, Yarmonite was considered a friend to white men. The present McCoy area was familiar ground to him. Yarmony Mountain, the park, the creek, the bridge and railroad siding are all named in his honor, as well as a neighborhood and street in Steamboat Springs. *Courtesy Tread of Pioneers Museum.*

from their ancestral lands and killed at will, weren't quite as willing to have strangers on their land. A few would mark the ground in front of James Crawford with messages such as "Ute's Dirt. Big Jim go back," according to Lulita Crawford in her book *Crawford Pioneer Tales*. Despite the young braves' threats, the elders were able to maintain peace for a time.

James Crawford brought his wife, Margaret, and their three children, Lulie, Logan and John, to settle. For several years, they maintained peace with the Utes. Come summer, the Utes would arrive with the men riding the horses they considered to be sacred. The squaws would walk behind them, leading horses and carrying their teepees, belongings and, somewhere in the mix, their children. One way the Crawfords kept peace was with a steady supply of biscuits covered with molasses or sugar. The Crawfords also provided medical service and went out of their way to keep relations

The Crawford clan. James Crawford founded Steamboat Springs in 1874 and then moved his wife and three children out to the valley. *Courtesy Tread of Pioneers Museum.*

peaceable. Friendships were even formed between Logan Crawford and Chief Yarmonite's son Charlie Yarmonite.

The first Fourth of July that the Crawfords stayed in the Yampa Valley, they were sent a bunting with red and white stripes, a blue field with thirty-seven stars, one large one in the middle, two circles of smaller stars and a star in each corner. When the Utes first saw the flag, they thought it was a declaration of white man's rule. James Crawford, seeing their discomfort, invited the Utes to the flag-raising celebration, where he declared that the flag was to watch over both red and white men, according to Lulita Crawford's *Crawford Pioneer Tales*. As the Crawfords tried to raise the flag, kinks in the rope hindered the raising. Chief Yarmonite's nephew was able to climb the flagpole and get the bunting flying. The Utes cheered and sang, and a fine celebration was had by all. For three summers, the Crawfords and the Utes coexisted. The Crawfords became as intimate with the landscape as the Utes were.

Lulita Crawford Pritchett writes of the beauty of the medicine or mineral springs as seen through the eyes of the Crawford children during their first years in the valley. From *The Cabin at Medicine Springs*:

> *In the other direction she* [Lulie Crawford] *could see pink clouds reflected in the pond formed by several of the mineral springs. The pond was the busiest spot in the valley these days. Ducks, muskrats, and beaver made trails across it. Snipe and rails skirted the shallows. Even now, at dusk, a clamor arose from the tangled reeds surrounding it, topped by the chortling of red-wintered blackbirds. Between the pond and the cabin, and along the base of the hill, there lay a rumpled sea of dark-green rushes, among which the springs glinted. Each spring had built a mound for itself, presided over by dozens of frogs that added their voices to the evening chorus…Bears came nightly to the mineral marshes to hunt frogs, their oozy tracks in the mud proclaimed. Deer had trailed up from winter range in the lower country, and elk were in all the groves, eating the juicy aspen twigs. The skyways had been filled with sandhill cranes and geese and ducks returning…All the game trails led to them as to a salt lick. Soon the grass fringing them, flavored with minerals, would be trampled and closely cropped. Now it was still fresh and green.*
>
> *She was thinking what good times she and Logan and John had had playing follow-the-leader among the labyrinth of pools. There were a few solid spots, but most of the knolls were spongy and jiggly. It took expert calculation to jump from mound to mound and not miss their footing. If they slipped, they sometimes sank to their waists. Lulie had discovered the Iron Spring by stepping in it, and now it was everybody's favorite. Ma always insisted they take off their shoes and stockings when they explored the marsh. Then, before they went to bed, she further insisted they scrub the iron-stained mud and the soda-white clay from their feet. John always fell asleep before he could get his feet washed, and Lulie invariably had to finish for him.*
>
> *The children knew every spring—the little ones that tinkled in their small, rush-grown cups, and the big ones that burbled loudly. All made a kind of friendly conversation. Lulie loved to watch the chains of iridescent bubbles rise to the surface and burst. She and her brothers examined the odd, jointed reeds and grasses and picked the mallows that bloomed in season. They squeezed the green blobs of algae and admired the pink scum on some of the pools. They considered the inhabitants of the marsh their special friends—the killdeer, frogs, muskrats; the red-winged blackbirds and shy, long-faced Wilson's snipes, the tiny songsters whose name they did not know.*

With more white people settling the Rocky Mountains, the U.S. government felt the need to manage the American Indians as encounters became more frequent and violence did too. The government began rounding up the Utes, removing them from their ancestral lands and placing them in reservations.

By 1879, the Utes had been relocated to the White River Indian Reservation in what is now known as Meeker, Colorado. The town today holds the name of Nathan Meeker, a missionary from the East Coast sent to educate the Indians. Nathan Meeker was ignorant of the American Indian way of life of migrating for food, hunting and enjoying life. Meeker is said to have been unsympathetic to the lifestyle of the Ute Indians and was determined to get them to give up their hunting and embrace an agricultural lifestyle. First, he tried to rein in their hunting expeditions by only handing out government rations once a week. Family members had to be present to receive rations. Then he took over their horse-racing grounds and plowed them for agriculture. For many Utes, this was the last insult they would stand. As tensions began rising among Meeker and the Utes, aggression also started to mount in Steamboat when surveyors arrived to officially designate Steamboat as a town.

James Crawford paid to have surveyors come and mark out the township of Steamboat. "Someday there'll be ranches all over this valley," he said. Lulie Crawford marked the visit from the surveyors as the moment the town went from being wild to civilized. "She felt a small pang that the beautiful unspoiled land would have to fit itself to the measurements the surveyors made. It would never again be free and wild. It would never again belong to the bear and elk. Fred said it would be labeled on a map *Township 6 North, Range 84 West* and this label would link it forever with the rest of the civilized United States," as retold by Lulita Crawford Pritchett in *The Cabin at Medicine Springs*.

A township is six miles square, but marking it off was no easy task. Fires had been burning throughout the summer, with smoke filling the valley. Some historians say the fires were set by the Utes, either in retaliation to their treatment by Meeker or to drive animals out for hunting. The Utes, on the other hand, pointed to the many fires started by the railroad workers as they cut ties. Also, the summer was a dry one, and the Utes also claimed the fires were started by lightning strikes. Utes used to burn trees to dry wood for the following years. While the fires were valuable to the Utes, white men prioritized live trees for lumber instead.

As the surveyors were working, Chief Colorow, who was known to have a more aggressive approach to the whites than Chief Yarmonite, confronted them with his son Piah, as recorded in Crawford's *The Cabin at Medicine Springs*:

> *"What you do here?" Piah demanded of the surveyors.*
>
> *"We're working for Mr. Crawford," Fred told him.*
>
> *With a haughty manner, Piah indicated the compass in the leather case.*

"You drive sticks in ground and chop trees. Make marks. Many white men come. Build fence. Take Ute country."

"Hey now, wait a minute!" cried Fred.

Piah gestured violently. "Utes have to go away. No more grass for ponies. No more ponies. No more hunt."

The Indians all had guns. The surveyors were unarmed except for Teddy, who had a rifle leaning against a tree. He was about to reach for it when Fred snapped, "Hold it!"

Not a breath of air stirred in the grove. Not a leaf moved.

Piah slid from his pony. With a long finger he scratched in the ground. "Ute's dirt!" he grunted. "One sleep, you go!"

Who is to say who was wiser, the elders or the young braves? Through peace, Chief Ouray still had to move his tribe off ancestral lands to the desolate environment of a Utah reservation where game and hunting opportunities were scarce because their first reservation in southern Colorado proved to be too valuable in minerals. White people continued to settle lands set aside for the American Indians. War and fabricated slights with American Indians were an easy excuse to push the Utes off their lands. Colorado newspapers often declared, "The Utes Have to Go."

That summer of fires, the Meeker Massacre occurred. Nathan Meeker had called in the army because he was afraid the Utes would rebel against him for plowing their horse-grazing grounds. Instead of working with the Utes or trying to understand their needs, Meeker simply tried to force them into what he thought was a godly way of living. The men who ran reservations were often missionaries sent by the church with the blessing of the U.S. government. The Utes knew soldiers were coming and went to meet them. They asked to settle the problem off the reservation. Due to a series of miscommunications, the soldiers were forced to take a different route to an agreed-upon meeting area and ended up on Ute land. The Utes felt they had been betrayed. War was waged, and many lives were lost. Meeker was killed, and his wife and daughter were taken hostage for a while by the Utes until Chief Ouray intervened and demanded the whites be let go and the fighting end.

As the violence erupted, messengers raced through the Yampa Valley with word of war. James Crawford was out of town on business. Mrs. Crawford and the children spent a fearful night north of their home in the woods before being advised that neighbors from far away had gathered at the Crawford cabin to form a resistance against any attack. There never were hostilities in Steamboat, aside from an incident where Chief Colorow's nephew, drunk from whiskey, mimicked scalping little Lulie's doll. However, the fear stayed with the whites for a long while.

In southern Colorado, the Utes reached a treaty with the United States government. They would be moved to a reservation in Utah with Chief Ouray's tribe and other tribes already established there. They were not permitted to stay in Colorado because the minerals found in the mountains had more value to the U.S. government than the lives of the American Indians.

As recorded in *The Cabin at Medicine Springs*, Lulie Crawford asked her father, "Will the Utes really have to leave the mountains forever?"

He responded, "That's what the newspapers have been hollering for a long time," frowning. "I guess it was bound to come sooner or later. Civilization is pushing into the last frontiers, and the Indians have to give way. It'll be easier for the homesteaders not to have the redskins bothering around, but I can see the Utes' side, too. These were their mountains first."

"Where will the Utes go, Pa?"

"Somewhere west into country no white man wants."

So it was that in 1881, the Utes were relocated to the Uintah Reservation, where the land was barren and where they were unable to hunt to survive during the long winters. Without hunting, they were dependent on government rations. In the fall, the Utes would leave the reservation and return to Colorado to hunt, killing thousands of deer and elk to support the tribe. Because of the Utes' hunting and the hunting by whites, for a time, elk hunting was no longer permitted in Colorado as numbers had declined so excessively.

As Colorado became more widely settled, hunting limits were put in place, and the Utes were found to be breaking these because they were unaware of them. Through more misunderstandings and the quick-draw nature of white troops, a few more battles ensued and the Utes were removed from their homelands forever. In 1905, the federal government withdrew 1,110,000 acres from the reservation to create the Uintah Forest Reserve. Many years later in 1933, the Utes received some $1,100 each in payment for this land. It was not until 1945 that 236,000 acres of land that had not been homesteaded by white people were restored to the reservation, reports Wilson Rockwell in *The Utes: A Forgotten People*.

The Crawfords were able to successfully claim the township of Steamboat Springs for their own. They did miss their dear friend Chief Yarmonite, whom they credited with their safety during the Meeker uprising. Crawford writes in *The Cabin at Medicine Springs*, "The November wind caught up the music and carried it across the mounds and rushes of old Medicine Springs, that would never again echo to war drum or medicine chant; and on across the snowy sagebrush, where someday would stretch the broad main street of Steamboat Springs."

In 1900, the town of Steamboat was incorporated. In 1902, people lived in the valley, which included four general stores, three banks, two dry goods stores, two newspapers, three livery stables and more. Steamboat was originally established as a dry town. Perhaps more than the Crawfords' religious beliefs on the subject, they had seen firsthand the negative effects on the Ute Indians. There was a liquor clause restriction set up that was included in property deeds. The clause stated that if liquor were sold from the property, the owner would forfeit the property. In opposition to the liquor restrictions, Brooklyn, the neighborhood across the Yampa River from downtown Steamboat, established bars and brothels and became known as the seedy side of town. The liquor restriction stayed in place until June 15, 1939. Mrs. Margaret Crawford's death notice was posted in the paper the same day as the lifting of the liquor ban.

The next step in creating a town was to attract people to come. Aside from a few people who spent a year here or there, the town was still largely undeveloped. James Crawford initially intended to build up the various springs into a spa-like area, but he knew the greatest boost to the population would come from the railroad. That advancement was a long time in coming.

After the Utes left the land, white settlers started to arrive in 1881. In 1882, S.M. French opened Steamboat's first store. Rumor has it he may have been drawn to Steamboat because of the reputation of the healing mineral springs. His son, Bertie, had fallen from a high chair and been left permanently crippled. Perhaps French sought a cure for his son or merely relief for his son's pain.

By 1888, the future of Steamboat had been predicted. The brochure "The Future Metropolis of Routt County, Colorado," printed in 1888 by the *Summit County Journal*, detailed why people should come and what to expect in this most incredible of valleys:

> *ROUTT COUNTY.*
> *The Garden Spot of Colorado.—Steamboat Springs its Metro-polis—The Coming Great Health Resort of the Union*
>
> *Among the many beautiful places of pleasurable and healthful residence and resort, throughout the mountain scenery and grandeur of the great and growing state of Colorado, Steamboat Springs with its natural and phenomenal wonders, its salubrious and temperate climate, its beauties in location and landscape, is destined to take its place among the most*

prominent and important. It is a romantic and blissful retreat surrounded with pleasurable objects and necessary advantages to promote health, comfort and contentment, a refuge for rest and retirement, a resort for pastime and pleasure, and a sanitarium for the weak and invalid. Such conditions and attractions will soon popularize among the public and the seekers will build a town,—a city in fact foremost among the healthful and grand resorts of the Rocky mountains or the west.

Its Location.

Steamboat Spring is located in the eastern part of Routt county on the Bear or Yampa river that flows north from Egeria park forming an elbow at this point; then flowing west through the length of the county. The town is platted on the outside of the elbow in a beautiful open valley or basin. The river hugging the abrupt but timbered and grassy cliffs on the inside. The valley is broad and open, with a sufficient level expanse for an extensive and convenient town site, but few feet above the river. A terraced bank rises then above the valley to an open plateau, presenting many lovely and sightly spots for homes and residences.

Water Privileges,

Three beautiful mountain streams break through the town plat—Soda creek, Crystal brook and Spring creek, springing from the summits of the western range, furnishing a bountiful supply of the most pure and wholesome water. The future advantages and utility of these streams can hardly be overestimated. The early morning sun peers through an opening of the eastern range, and sets in a line of the Yampa valley lessening the dreary effects of morning and evening shadows, so frequent in many mountain localities, and from the gradual rise, of the mountain ranges, with their slopes covered with luxurient vegetation and verdure, that excessive heat caused by refraction and reflection, does not occur in the middle of a summer's day, so noticeable and objectionable in many mountain places—No scorching furnace of a noon-day sun, and narrow canon mid the mountains bare, no blighting shadows that our natures shun, to chill the morning and the evening air.

Special Features.

The attractions at Steamboat Springs at present are its natural wonders and its enchanting scenery. In the vicinity there is also much to learn. Nature has established a library here of primitive wonders. Extinct races have left their traces among the archives yet enduring with the imprints of time. Mechanical improvements and artificial comforts are needed to perfect the order of things, and improve the existence of a natural labratory, and construct a place fitted for the use and enjoyment of man. The first to interest themselves in the value and importance of these springs was Hon. J.H. Crawford, present assemblyman from this district, heading a party of Missourians on a prospecting tour in July 1874. The Indians then occupied the whole country now known as the western slope and but few white men had ventured into Lo's dominions. Mr. Crawford with others of the party located here and have since remained with full confidence in the future growth of the country. There are palpable and unmistakable evidences that white men visited here as early as 1835. The remains of an adobe house built in the form and location of a fort, with portholes and surrounded with earth work approaches, are yet distinctively traceable, and a number of utensils have been unearthed and found about the premises; a logchain, oxshoes, nails, irons for a cart or wagon, &c. Trees about, long since dead, bear the initials J.D.—O.M.—C.H.L. with dates 1835 marked thereon when the trees were green, which are partially grown over. It is evident that a party of white men, explorers or adventurers remained here for some time in 1835, who they were or where they came from will likely remain a mystery. The attractions of the country most likely prompted them to remain sometime among the western wilds and savages.

Again in 1860, Charles Utter, a famous Colorado Indian trader and trapper, now said to be living in Mexico, bivoucked about the springs. In 1864, Major Oaks noted in Colorado established an Indian agency here. It was abandoned a year later. Rude pictographs are found on some of the projecting and uncovered rocks, remote and lingering symbols of the extinct aborigines—mound builders or aztecs. The evidences are numerous that those "children of nature" established a mecca here as a resort for health or pleasure. The unappreciative Ute yet speaks of these healing waters as "Heap wano" or "Heap big medicine." Their attractions and benefits are appreciated by the barbarous as well as the civilized: These healing waters by symbols we learn, the moundbuilders drank from their plastic urn, they healed the Ute warrior, cured the pains in his maw, and healthfully painted the cheeks of his squaw.

NUMBER AND VARIETY.

The springs number over forty and vary in temperature from 156 degrees Fer. to nearly ice cold and each spring different in appearance and taste. Some of them are very palatable, delicious and refreshing, especially the soda, iron and magnesia springs. No artificial mixture compares with them as a pleasing beverage. By chemical analysis these waters are found to contain potash, soda, magnesia, lime, sodium, iron, silicic acid, phosphoric acid, alumnia, manganese oxide, impregnated with many kinds of gasses.

Their curative and healing powers are undoubted. For rheumatism, urinary and cutaneous diseases, they seemingly act as specifics. The baths are strengthening and exhilarating and do not produce that enervating condition of the system often produced by thermal mineral waters. The projected vapor baths will yet constitute an important feature in hydropathic treatment. Some trouble and expense are required to put the natural openings of the earth in order that contain these mineral vapors. Some of these springs discharge a vast amount of water with never varying quantity. The heated waters melt away the snow and ice in their course to the Yampa and flowers bloom about the springs in winter. Strange and phenomenal are some of these springs. A prominent iron spring after the earthquake of 1882 discharged for a year afterwards a blackish substance probably caused by a displacement of the iron furniture in the chemical labratory below. One large spring presents a basin at the surface, with a mineral formed rim, about a rod in diameter filled with a mineral solution, the exact resemblance to milk.

For ages upon ages its sound has broken the solitude of the wilderness and beat the measures in the cycles of time. It seems like the exhaust pipe of wonderful and mighty engines that are working below, all in all for extent, for wonder, for variety, for study, for present known uses and their probable utilization in the future, these mineral springs are not surpassed on this continent, and what makes them still more important are their grand and matchless surroundings. The beautiful drives, that extend for a score of miles in different ways, the pleasant walks, where lovers are wont to tread, the rippling streams, cascades and pools, the sylvan bowers, the glens, the birds, the fish, the flowers, nature animate and inanimate, all tend to create an inspiration of joyous thought and a consciousness of earth's loveliness, not produced in every clime.

Preference Given to Settlers.

During the past winter a demand for lots has been made by speculators, coming from every direction, but the company prefer to give the actual settler the preference. The instant that a railroad points this way a boom of no small dimensions, will be inaugurated, capital must go by steam and when it is ushered in, grand projects will be made known, will turn to interest and the world will be better apprised of the northwest section, just over the range. An effort will soon be made to open a highway to some point on the Midland road on the Eagle river. The distance will be less than sixty miles and a good road. By fall 3,000 people will receive their merchandise via Leadville and the Eagle. Dillon and Rawlins will have to then seek new kingdoms of trade. Sixty miles to a railroad, a year ago it was 120 miles.

One more turn gentlemen, and the gap will be closed. When it is closed our resources will begin to expand and Steamboat Springs will soon become a city and the principal public resort of west.

By 1899, the town's first newspaper went to print. The *Pilot*, today known as the *Steamboat Pilot & Today*, published a special spring edition. The title read, "The Largest Group of Mineral Springs in the World Is at Steamboat Springs." "One hundred and fifty natural fountains giving forth their curative waters are found here. Their constituents vary as greatly as their temperature. The latter ranges from ice cold to 1,500F. The combined discharge of the group is 2,000 gallons per minute." For many, the springs were the primary reason to build a life in this area.

In addition to the mineral springs, another draw for tourists to Steamboat was the annual Fish and Game Day. "Colorado Notes" in the *Akron Weekly Pioneer Press* of Friday, August 3, 1900, reported:

Game and Fish day will be celebrated at Steamboat Springs September 13th and 14th. The executive committee in charge of the celebration is composed of James II. Crawford. Chairman; C.H. Leckenby, secretary; J.N. McWilliams, F.A. Metcalf and B.L. Jefferson. They have succeeded in raising by subscription a large amount of money and fully $1,000 will be expended in giving the visitors a good time. The principal feature, of course, will be the wild game barbecue, and the arrangements are being made to make this feature even more successful than last year. The hunting will be in the charge of the Steamboat Springs Gun Club, in which are some of the crackshots of the state. The barbecue will consist of bear, deer, elk, feathered game and fish. Preparations will be made to feed 5,000 people. The gathering of the meat will be a big job, but will be done in strict accordance with the law. Outside sportsmen will be utilized at least such as wish to join in the hunt under direction of local guides.

Even though the Fish and Game Days were enticing, the springs still had their pull. One of the earliest brochures written about the springs by a well-known physician of the time stated:

Mineral springs are magical in their charm for the layman. All people have an abiding faith in Nature and the mysterious workings that are constantly taking place in her unseen laboratories deep down in the formations of the earth, for they feel she is dispensing chemical combinations far beyond the possibility of man to do…For such an asset cities would give thousands of dollars, yet strange to relate, not five % of the people in Colorado are aware that we have that very asset in our midst.

The springs finally started to gain quite a bit of attention in newspapers from Steamboat to Denver. As word spread of the valley's bounty, talk about a railroad passing through finally came to fruition.

In 1902, David H. Moffat signed a contract to begin building the Denver Northwestern & Pacific Railroad that would pass through Steamboat. Although the contract was signed, it wasn't until 1908 that the railroad finally reached Steamboat. At one point, the railway company threatened to bypass Steamboat altogether for a cheaper, easier route. However, the citizens of Steamboat were determined to have the train come through their town. They raised a cash bonus of $15,075 for the railroad. The Steamboat Springs Company, in part owned by James Crawford, donated one hundred acres and $1,000 to build the brick depot that still stands today. Today, the trains merely take coal from town to town instead of passengers. An idle train car sitting on cement next to the Art Depot reminds visitors of the excitement the townspeople must have felt when "civilization" finally arrived in their midst.

At the same time the railroad negotiations were taking place, those in Denver were taking interest in this little town tucked in the valley. The following, written on June 8, 1905, is from the collection of the Steamboat Springs High School Library; the author is unknown:

Whence came these legends and traditions?
With the odors of the forest.
With the perfumes of the valley,
With the curling smoke of wigwams
With the song of peaceful rivers.

From the mountain, moor and fen-land,
From the lodges of the beaver
From the hoof-prints of the bison,
From the eyrie of the eagle
From the portals of the sunset.

At the eastern end of Routt County in North-west Colorado, where the Great Divide in its majestic trend across the continent, subsides to an elevation

The original train depot was paid for by James Crawford in order to entice travelers to Steamboat. Today, the depot is used by the Arts Council to display the artwork of local sculptors, painters and photographers. © *Dagny McKinley 2012.*

some distance below timber line, then into undulating foot-hills and gentle valleys, at the great bend of the beautiful Yampa (Bear) River and at an elevation of 6,500 feet, are situated Steamboat Springs and the town by that name. The history of this region very soon enters the fragmentary confines of unwritten talks and then vanishes into fable. Civilization, in its westward progress toward the Golden Gate passed north, south and around it, leaving unmolested to the dominion of the Indians, animals and birds, these streams, valleys and forest. Practically this primitive life and condition existed until about 1835, when were noticed the hand prints of the ubiquitous and invincible Jesuit missionary. Just out of the view at the right-hand edge of the picture of the town of Steamboat Springs at the foot of a giant cottonwood he built his adobe hut. Some of the blocks of this structure to this day show the shape of his fingers where they were grasped to be placed in position. The ruins of the house are yet to be seen beside the dead and prostrate form of the old tree. Both of which will be carefully preserved as long as human vision can distinguish their outlines.

Here was the interior, the holy of holies of the Ute kingdom. The enemies of this clan of aborigines, to reach it must conquer many natural difficulties, must cross the American Alps. Nature, in fact, to a great extent, fought their battles and vanquished their foes. Finally, however, a few years subsequent to 1835, a mighty host of Cheyennes and Arapahoes, under the leadership, it is said, of a white chieftain, crossed the mountains and advanced upon this place, where at that time was concentrated a large part of the Ute nation. For days the sanguinary conflict raged. Over the surface and slightly buried in the soft soil are found in great numbers, symmetrically fashioned arrow heads of chalcedony and brown jasper, strewn here most, if not all of them, as a sequence of that tragic episode. A small flat-topped eminence of a few acres in extent, situated a half mile westerly from the Pavilion Spring, and skirted by a grove of fine old cottonwoods is said to have been the culminating point of the battle. Arrow and spear points are here found in such quantities that is called "Arrow Head Mesa." So fierce and fatal, was this meeting, to the three tribes involved, that from that time they have been shy of each other and loath to begin war.

Intrigue remained, and people came. With the railroad making stops in Steamboat, "outsiders" started to look at the financial possibilities of developing Steamboat. The Steamboat Springs Town Improvement Co. was formed and a list of projects compiled that included (according to the *Steamboat Pilot*):

- *Build first class modern hotel—about 100 bed rooms*
- *Build handsome and modern bathhouse to replace old one*
- *Build bottling works to bottle, ship and sell the carbonated waters*
- *Set up precipitating tanks to save sulphur from the sulphur spring*
- *Pipe hot water from the springs for heating purposes (sell heat). It was thought the flow of water could heat a town of 20,000*
- *Estimated annual income—$58,600*

Unfortunately, of all the projects considered, only a couple came to fruition. One was the building of a new bathhouse that showcased private pools, a warm indoor swimming pool, an upstairs dressing room and a lounge and outdoor swimming pool. No bottling works were created until H.W. Gossard came to town in 1931.

The second was a modern hotel, the Cabin Hotel, built in 1909. The hotel was host to travelers near and far. A letter remembers that era:

> *Dear Bertha,*
>
> *Yes, you remember the "Cabin Hotel" correctly as the beautiful two-story, 100-room, resort complex at the west edge of town surrounded by a variety of mineral springs. The hotel was built soon after the railroad reached Steamboat (which was in 1909). Well, actually the rails reached Steamboat in Dec. 1908, but service didn't commence until the following spring.*
>
> *The hotel was located on the north bank of the Yampa River, right across the river from the Moffat Depot.*
>
> *The Main Street of Steamboat is a perfectly straight one-mile segment of U.S. 40. The Cabin Hotel was at the west end of the main street, Lincoln Avenue. The eastern terminus of main street is at the Heart Spring, a large hot spring that supplies indoor pool and private baths known as the "Bath House," and a large outdoor pool.*
>
> *Prospective guests and other railroad passengers were met at the Depot by a "Concord Coach" provided by the hotel to transport travelers and their luggage over town. And on occasion, this coach is used to convey hotel guests to and from the Bath House.*
>
> *It is said that there are over 100 mineral springs within a quarter-mile radius of the hotel, with the Heart Spring only a short distance outside this circle. I'll mention only a few of the most prominent springs. On the north side of the river, and no more than a few roads from the hotel, some of the best-known springs include the Sulphur Spring, the Iron Spring and the Chinking-Mud Spring. On the south side of the river, within a few roads of the Depot) are the Lithia Spring and the Steamboat Springs. The Steamboat Springs are a pair of small geysers that erupt alternatively from opposite sides of a common basin, a few feet in diameter.*
>
> *Well, actually, I am describing the Steamboat Springs as they appeared to James Crawford, the founder of the town, when he staked his claim here in 1874. In those days the geysers sprouted water with a rhythmic "chugging" which sounded for all the world like the old Missouri steamboats.*
>
> *Although a generous flow of water can still be seen bubbling from each of the two springs they no longer spurt water into the air, as in the old days.*
>
> *Local authorities blame the railroad builders for destruction of this remarkable phenomenon.*
>
> *The geyser effect of the springs, however, had greatly diminished prior to the railroad's arrival possibly from natural causes, possibly on account of*

Heart Spring stained glass that hangs in the window of the Old Town Hot Springs. © *Dagny McKinley 2012*.

youngsters chucking stones as deep as they could into the springs, to see how high the geyser would throw the stones.

Meant to write more about the hotel. Well, maybe next time.

Until then, sincerely,
Anna

The Cabin Hotel burned down but was once a grand sign of progress overlooking the valley from where the Bud Werner Memorial Library sits today.

In 1910, another call for permanent residents went out to potential settlers in the form of a brochure titled "Steamboat Springs & Routt County Colorado brochure," compiled and published by the Northwestern Loan & Investment Company and edited by James Logan Crawford:

> *A century ago when the northwest corner of Colorado, now known to the world as Routt county, was but a portion of the unexplored Spanish America, Jesuit missionaries at the missions along the Gulf had heard stories from the Indians of wonderful springs of water far to the north presided over by Manitou himself, who breathed through the waters and gave them the power to drive away disease. These stories seemed legendary to the good Fathers and were undoubtedly considerably embellished by the Indian narrators, but from the descriptions given of the beautiful valley and the talking spring, there is little doubt but that the now famous Bear River valley and the Steamboat Spring were the basis of the legends.*
>
> *No more picturesque location for a town could be conceived. Nestled at the foot of gently sloping verdure-clad hills on the banks of the Yampa or Bear river, with the main range of the continental divide forming the background to the east, and timber-clad, undulating hills across the river to the west. From the south comes the Bear river and the valley through which it wends its way, once a tangle of jungle and forest, is now filled with prosperous, well cultivated farms. Just east of the town, the river makes an abrupt turn to the west and from the town can be had an extended view of the valley to the west and to the south, while over the low hills to the north can be seen the gentle slopes of Elk Mountain in the adjoining valley. No wonder the Indians thought that God had made this location for His own earthly resting place and there is plenty of evidence that the wonderful springs of this locality were patronized by the Indians for centuries before the advent of the white man and his civilization.*
>
> *Steamboat Springs is already the metropolis and commercial center of Routt county, the largest county in Colorado, an empire in itself. Distant but 175 miles from Denver on the new Denver, Northwestern & Pacific railroad, locally known as the "Moffat Road," at an altitude of about 6,500 feet above sea level, it occupies a position as the gateway to Routt county, and with its mineral wealth of coal, gold, silver and Copper at the door and natural springs of wonderful medicinal value, its position as the largest and most important city between Denver and Salt Lake is assured beyond question.*

In summer the climate is perfect; the days being warm and delightful; the nights cool and refreshing and entirely without the humidity and sultriness experienced in eastern localities through excessive heat. In winter the weather is fine and bracing; the majority of days clear and bright. There is very little cloudy weather. The fall of snow is heavy in winter, but there are few storms and practically no wind. For about three months there is the finest sleighing and snowshoeing to be found anywhere and, as a winter resort, Steamboat Springs presents every attraction for those seeking health and recreation. It is a splendid outfitting point for tourists and sportsmen, being in the center of the greatest hunting and fishing section remaining on the continent. For pleasure seekers it offers every attraction imaginable. The State Fish Hatchery is located here and annually places over one million trout in the streams of the county. Three mountain streams course through the town, including the beautiful Bear river, which traverses the entire county from east to west.

Steamboat Springs is favorably situated with regard to the various resources of the county. On the east and north are the precious mineral deposits; a short distance to the west begin the great coal fields, while the ranches of the surrounding valleys south and west produce abundantly of the products of the soil. Building stone, marble, onyx and lime lie in inexhaustible quantities at the very door. A planing mill, lumber yards, brick manufacturing plant, flouring mill, cigar factory, steam and hand laundries, three hotels, three banks, two newspapers, Commercial club, large, up-to-date stores carrying full lines of the various branches of merchandise and everything needed by the ranch man, tourist, traveler or resident, all form part of the town's business life. In a word, Steamboat Springs has greater promise of the future than any other town in Colorado.

…The mineral wealth of Steamboat Springs and Routt county simply cannot be determined. The amount of the coal supply and the extent of the coal fields alone surpass all comprehension. The Yampa coal field itself, according to government engineers, embraces approximately 1,200 square miles. This basin begins very near Steamboat Springs and covers the Oak creek and Trout creek field. It takes in the Pilot Knob or anthracite coal (Hayden and Williams Fork) and extends down the Bear river in a due westerly course to Craig. The field west of Craig is fully as large as the Yampa field, and takes in the Lay, Maybell and Lily park districts. The two fields combined make an area of nearly 2,000 square miles, or about a million acres, being over 100 miles long and about 20 miles wide.

> *...Eminent geologists have pronounced Routt county a great oil basin where with proper development a rich industry will be established in the production of oil. There are oil springs where crude petroleum flows from the ground.*
>
> *Routt county might well be styled the "New Canaan," for it is truly a land of promise. Volumes could be written concerning her great and unlimited resources without the slightest exaggeration. It is indeed a land of opportunity. There are more favorable opportunities that can be taken advantage of by settlers and investors in Routt county than in any other section of the west. In commerce, irrigation, mining, manufacturing, coal, oil, agriculture, stock raising and many other pursuits there are abundant opportunities for accumulating a fortune. With the Moffat road spanning the county from east to west it will leap into productiveness and wealth. Today it is the land of opportunity and now is the time to embrace that opportunity. For scenic beauty the Moffat road has no equal even in the famed Alps or Andes and for engineering skill and railroad construction stands a marvel of modern times. A trip over this road and a visit to Steamboat Springs will be an experience ever pleasantly remembered, and the man who has once feasted his eyes upon the hilts, the valleys and streams, felt the spell of the mighty mountains, and breathed the fresh, pure, life-giving air is sure to find his way back to this country. The Moffat road opens up to settlement a new territory in northwestern Colorado and northeastern Utah larger than the six New England states and New Jersey combined and capable of sustaining a population of at least 5,000,000 inhabitants.*
>
> *The eyes of the industrial world are upon the west and the construction of the Denver, Northwestern & Pacific railroad is pointing the way through Routt county as a vast and undeveloped empire inviting the judicious expenditure of energy, brains and capital.*

People arriving meant industry, and the main industries in the early days were ranching and coal mining. In the five years following the arrival of the railroad, more cattle were taken to market from Steamboat Springs than anywhere else in the United States.

In 1913, the course of Steamboat's future changed forever. Local resident Marjorie Perry was in Denver as a spectator to Carl Howelsen's ski-jumping show. It was Marjorie who convinced Howelsen to visit Steamboat. Not only did he visit, but he also built the town's first ski jump. The original exhibition was the impetus for the very first Winter Carnival festival that continues to this day.

The Gossard family, circa 1925. *Left to right*: H.W. Gossard, Bill, (man in back is unknown), Gloria and Pearl Gossard. The family is sitting at the Sulphur Spring. Lincoln Avenue is directly behind them, and above the stone wall is Iron Spring. *Courtesy Tread of Pioneers Museum.*

Howelsen not only introduced ski jumping but also encouraged skiing as a recreational activity. Until that time, skiing was used as a means of transportation in winter. His influence on Steamboat until his departure in 1921 cannot be underestimated. Steamboat slowly changed from a destination spot for healing waters to a destination ski resort.

While skiing was taking off, the springs remained a place for relaxation and inspiration. However, they had not yet achieved their fullest potential as a tourist attraction. In 1931, H.W. Gossard thought he might change that. From French Link, Indiana, Gossard had recently retired from his corset business. The city leased the town's springs and Howelsen Park to him from 1931 to 1935. In order to better market the springs, he changed the name of the Bath House Spring (Old Town Hot Springs) to Heart Spring and built a wall of stones in the shape of a heart around it. He labeled the Heart Spring recreation area the Rocky Mountain Miraquelle Spa. In order to promote his new venture, he created brochures and was the first on record to analyze the mineral content of the most important springs. He had deep faith that the Lithia Spring (otherwise known as the Milk Spring) and Soda Spring would bring people from far and wide. His dreams were not realized, but his legacy remains because his work resulted in a community gathering place at the Heart Spring that remains a focal point for community and visitors alike to this day. Gossard died at the ripe old age of ninety-four in 1965.

When 1940 arrived, there were still high hopes that Steamboat's mineral waters would become one of the most famous tourist attractions in the world and able to bring in millions of dollars a year. An article in the *Steamboat Pilot* claimed, "The Steamboat Springs group of mineral waters is the largest and most varied in the world and there is found here nearly, if not quite, every water known to have potent value. Alkaline, semi-alkaline, iron, Lithia, sulphur, magnesium, in all their various and wonderful combinations are found here—more than 150 springs in all, and none the exact counterpart of another." This was only disproved many years later with the study of the Colorado Geological Survey that found the Old Town Hot Springs and Strawberry Park Hot Springs do indeed come from the same waters.

"Rheumatism, gout, scrofula, liver, kidney, dyspepsia, the virulent blood disorders, skin diseases and anaemical diseases having their origin in an excess of uric acid in the blood, and many other complaints of the human organism find ready relief by the use of these waters," continued the article.

The springs were compared with the famous Kreuznacher in Rhenist Prussia and the water of Bath, Vichy, Flombleres and Baden-Baden for their radioactive qualities similar to the Hot Springs of Arkansas. The abundance of the flow of the springs also made them attractive; they are estimated to have a combined flow of two thousand gallons per minute. The story concluded, "Nothing in nature is lacking to make it the greatest health resort in the world."

"Mineral Springs Have Great Potential Value" touted another newspaper article:

> *The drinking spring—just one, mind you—near Corblenz, on the Rhine in Germany, from which the world famous Appollinaris water is bottled has been the property of the Rothschild family, Europe's greatest family, for more than a hundred years. While it is impossible to obtain any detailed information from its owners concerning the revenue from the great property it is popularly reputed to have received a net profit of $2,000,000 each year. Aside from the tourism dollars famous springs draw in, water is often bottled and sold and the mineral salts that come from an evaporation process were used in many pharmacies.*

As the eager author of the article noted, natural springs have the potential to create revenue until the end of time unless they are disrupted or the fault lines under them shift, causing springs to disappear. In Wisconsin, a well-known spring sold for $1 million in the early 1900s. In Colorado Springs, the Manitou soda springs were sold for $250,000. Visitors to the Manitou soda springs once numbered over 400,000 per season, spending close to $8 million.

Today, both of Steamboat's hot springs are revenue earners for private industries. The Steamboat Springs Health and Recreation Association (SSHRA) is a nonprofit organization set up to maintain the Old Town Hot Springs for the community's benefit. It puts profits into expanding and rejuvenating the property for the best visiting experience. The Strawberry Park Hot Springs are owned by a private individual who has cleaned them up and developed them in order to make them a destination for out-of-town guests and locals alike. Steamboat's Soda Spring has disappeared. Why Steamboat's springs never achieved the fame of some other of the world's most famous springs is hard to say.

By the 1950s, automobiles had surpassed the train as a means of travel. Visitors were greeted on a trip into town with the effervescent waters of the Soda Spring. Fizzy lemonade was made by simply adding fresh lemon. The gentlemen of town were more than happy to prepare such a beverage for a pretty lady on a hot day. Different flavors of soda were experimented with, and talk came and went of bottling the carbonated waters that sprang out of the ground. But talk is all it ever amounted to.

Visitors would often carry away gallons of medicinal water. Mr. Gossard, one-time owner of the Lithia and Heart Springs, was known to have a tank in his car that would safely hold the Lithia waters of his "beloved spring" so that he would never be without them. People from Denver would request to have water shipped to them for personal use.

Steamboat continued to grow. Skiing became the town's top income earner. Springs began to disappear. Some were claimed by construction and others trampled by herds of tens of thousands of cattle that used to roam the valley. According to the *Routt County Sentinel*, at the end of 1902, there were an estimated 100,000 head of cattle and 20,000 head of horses in Routt and Moffat Counties. Some mineral springs were filled in with rock, and some were lost by not learning a lesson the first time. Today, the springs that are still alive are preserved with signage and concrete viewing platforms. The areas have gone from eras of disrepair to an era when they are cared for and available to the public. People are welcome to drink from the springs, although bacteria in the water is a possibility. To appreciate the different flavors and to identify the mineral content of the waters, taste is the best measure.

Local resident George Tolles continues to help educate people about the various springs. He penned the guide "A Walking Tour of the Springs of

Steamboat" and is responsible for the signs that mark each spring. He also added memorable quotes that nod to the high regard water and mineral springs have had throughout history from noted authors such as Thoreau, Wordsworth and more. As Tolles notes in an interview with the *Steamboat Pilot & Today*, "The mineral springs, which are not all hot springs, [they are] part of Steamboat's heritage that surpasses even skiing." Even now, Tolles drinks Lithia Spring water twice a day.

Whether the healing properties believed to generate from these springs come from the minerals in the water or the relaxation that floating in the waters provides, one cannot be sure, but with close to fifty thousand people a year enjoying the Old Town Hot Springs alone, water is surely still a draw.

Chapter 2

Black Sulphur Spring

Mineralogy of Spring, 2012

Temperature	70 degrees
Alkalinity, mg/l	2218
Bicarbonate, mg/l	2218
Boron	nonc
Calcium, dissolved, mg/l	49
Carbonate, mg/l	none
Chloride, mg/l	1230
Coliforms, -/100 ml (1/100 ml)	none
Fluoride, mg/l (4 mg/l)	13.3
Hardness, mg/l	237
Magnesium, dissolved, mg/l	28
Nitrate, dissolved, mg/l	none
Nitrate/Nitrate, mg/l	none
Nitrite, dissolved, mg/l	none
Nitrogen, mg/l	none
Ph, units	7.8

Potassium, dissolved, mg/l (5,600 mg/l)	110
Sodium absorption ration	55.41
Sodium, dissolved, mg/l (3,000 mg/l)	1940
Silica, mg/l	none
Sulfate, mg/l (250 mg/l)	574
Arsenic, dissolved, mg/l (0.05 mg/l)	.027
Barium, dissolved, mg/l (1.0 mg/l)	.03
Cadmium, dissolved, mg/ (0.010 mg/l)	none
Iron, dissolved, mg/1 (0.3 mg/l)	.39
Lithium, dissolved, mg/l	2.75

mg/l = milligrams per liter
Elements: Maximum recommended by state and federal government in ()
*Source: Tread of Pioneers Museum, "The Springs of Steamboat Springs Walking Tour"

Location

From downtown Steamboat, head southwest on 13th Street, which can be identified by the Bud Werner Memorial Library on the corner where the Cabin Hotel once welcomed visitors to Steamboat. Follow 13th Street across the bridge. Immediately after the bridge, you'll find parking on the right-hand side at the Art Depot. Black Sulphur Spring is on the opposite side of the street from the parking lot, or the east side of the street. This spring is marked with a sign and sits just above the Steamboat Spring and the Yampa River.

What to Expect

Black Sulphur Spring, or Ink Spring, is the spring whose waters are slate or gravel colored. The waters gently bubble and boil in their murky depths. While the dark coloring may look ominous to some, it is produced by hydrogen sulfide being reduced to sulphur. Visitors will not see their reflection in these dark waters, but intrigue as to what lies below will certainly be found.

The proximity of the railroad tracks to the spring shows how easily water flows were disrupted when the hillside behind the railroad was blasted. © *Dagny McKinley 2012.*

Early in the morning, between 6:30 and 7:30 a.m., a train usually passes through town, carrying coal from the mines. The trains once carried passengers from Denver to Steamboat for outings but now are used only for transportation of goods. Standing at Black Sulphur Spring as the train passes, one can feel the vibrations of the behemoth. So close is the proximity of the railroad to the springs that one could literally reach out and touch the passing freight. Some may be surprised these springs didn't disappear when the rock walls in the hill above the springs were blasted to make way for the railroad to come through.

People gather in the summer and crowd around the springs, not to ponder the beauty of the waters, but to enjoy such events as the Yampa River Kayaking Festival and the Rubber Ducky Race. Early morning and midday are picturesque times to visit these springs to feel the warm Steamboat sun and to imagine what life would have been like as one of the first settlers of this town when the hillsides were covered in green instead of houses and civilization was a memory away.

HISTORY

Not much has been written about this spring, possibly because its benefits for humans and the environment are as yet unknown. The uniqueness of this spring cannot be denied. From 1839 to 1874, Black Sulphur Spring was called the Ink Spring and was part of the Crawfords'—the first permanent settlers in the area—land claim. There was another spring in the vicinity known for its inky purple color, and a little was written on that; however, the two do not appear to be the same spring.

The following is courtesy of the Steamboat Springs High School Library:

> *PURPLE SPRING*
> *This spring is so named from a peculiarity not yet explained by the chemist. Its water varies in color from a slight to a distinct purple. Another curious feature it has is of periodically emitting sulphuretted hydrogen. It has a local reputation of being highly purgative. The flow is* [also] *accompanied by an abundance of carbonic acid gas. The analysis shows a water of rare value, both drinking and bathing, containing a small amount of lime and so rich in sulphur that there is an abundant precipitation of this mineral in the brook that runs away. One of the results of bathing in this water, is the skin of the person who is otherwise in reasonable health becomes soft and smooth. No artificial sulphur bath will attain, so perfectly, such an end. This, if anything will restore a harsh, dry, and muddy skin to a normal state and with the aid of the variously graded iron water near by, remove the splotchy yellow and replace with clear color and damask bloom. It is a prophecy that every country, and class, will send here thousands of devotees who will bow down in appreciation of its unexceptional merit.*

Chapter 3

Crawford Spring

Mineralogy of Spring, Early 1900s

Class of Water—Sodic, bicarbonated, muriated, alkaline-saline (carbon-dioxated, sulphuretted)

Constituents	**Approximately Parts Per Million**	**Reacting Value Percentage**
Silica	29.9	
Sulphate	546.3	5.59
Bicarbonate	3,267.1	26.27
Carbonate		
Phosphate		
Chloride	1,310.8	18.14
Iron		.03
Aluminum		

Constituents	**Approximately Parts Per Million**	**Reacting Value Percentage**
Iron Oxide		
Aluminum Oxide	2.75	
Manganeseone		
Calcium	130.66	3.2
Magnesium	35.83	1.45
Potassium	124.5	1.57
Sodium	2,051	43.75
Lithium	Trace	
	7,498.84	100.00

Location

The Crawford Spring is located just up the hill from the Iron Spring, across Highway 40 from the Bud Werner Memorial Library at 13th Street. Access can also be had by turning right on 12th Street (coming from the mountain) and turning left almost immediately into a parking area for the springs.

What to Expect

This spring runs year-round and was named the Crawford Spring because it sat behind the original Crawford cabin. According to Jim Crawford, descendant of the Crawfords who first settled the town of Steamboat, the cabin was actually four cabins forming a cross around a center courtyard; the architect's building is about where the south cabin was located, with the other three cabins extending east, west and north, so the back of the north cabin would be close to the Crawford Spring. The spring sits underneath a grand tree that offers shade in the summer or a perfect place for climbing. Elk can be seen traversing the hill behind the spring in winter.

History

There has been some confusion about which story referred to which spring, as the various springs' names have been interchangeable at times. The Soda Spring or Pavilion Spring was referred to in the first issue of the *Steamboat Pilot* (July 31, 1885) as the Crawford Spring. The Soda Spring was also referred to as the Iron Spring even though the Iron Spring sits adjacent to the Crawford Spring. There is no signage at the Crawford Spring. Instead, it sits like a treasure waiting to be found, just as Steamboat was James Crawford's treasure.

The Crawford Spring once sat in the backyard of the Crawfords' home. Today, the spring attracts elk and other wildlife. © *Dagny McKinley 2012.*

Chapter 4

Iron Spring

Mineralogy of Spring, 2012

Temperature	73 degrees
Alkalinity, mg/l	1644
Bicarbonate, mg/l	1644
Boron	none
Calcium, dissolved, mg/l	79
Carbonate, mg/l	none
Chloride, mg/l	438
Coliforms, -/100 ml (1/100 ml)	.6
Fluoride, mg/l (4 mg/l)	1.0
Hardness, mg/l	464
Magnesium, dissolved, mg/l	65
Nitrate, dissolved, mg/l	.07
Nitrate/Nitrate, mg/l	.07
Nitrite, dissolved, mg/l	none
Nitrogen, mg/l	.04
Ph, units	7.6
Potassium, dissolved, mg/l (5,600 mg/l)	60
Sodium absorption ration	16.57
Sodium, dissolved, mg/l (3,000 mg/l)	811

Silica, mg/l	none
Sulfate, mg/l (250 mg/l)	228
Arsenic, dissolved, mg/l (0.05 mg/l)	.001
Barium, dissolved, mg/l (1.0 mg/l)	.04
Cadmium, dissolved, mg/ (0.010 mg/l)	none
Iron, dissolved, mg/1 (0.3 mg/l)	.79
Lithium, dissolved, mg/l	.96

mg/l = milligrams per liter
Elements: Maximum recommended by state and federal government in ()
*Source: Tread of Pioneers Museum, "The Springs of Steamboat Springs Walking Tour"

Location

The turnoff to the Iron Spring is easy to miss unless you are looking for it. As you leave downtown Steamboat, keep your eyes open for an entrance just past 12th Street across from the Bud Werner Memorial Library and the West Lincoln Park on the east side of Lincoln. The spring sits just below a hill in what was once the backyard of James and Margaret Crawford's original home.

What to Expect

When I first arrived at the Iron Spring, I thought, "This is it?" These were the first springs outside the Strawberry Park Hot Springs that I visited. Compared to the hot springs, this was just a trickle in the ground. The spring is located next to a building that houses a few shops. The hill behind the Iron Spring is grassy and said to swell with water in the springtime. Just walking up to the spring, my feet were covered in mud as I slushed through the wet grounds surrounding it. The grass around the spring was tinged with iron. What was once a spring that quenched the thirst of Steamboat's first family and sat in their backyard is now a little hole in the earth with a few bubbles. Stonework surrounds the spring while runoff trickles down to the sign that explains the spring. The earth under the spring that populates this area is reported to be pockmarked with holes and tunnels of flowing water, according to workmen for the highway. I decided to sit with the springs for a while and take in what they have to offer now. I came away feeling this was a quiet spring, really beautiful, with its rusty red waters and not many visitors for most of the year. The first settlers called this site home, and the spring seems to carry a memory of former times as they bubble out of sight of most passersby, offering a glimpse into history for those who choose to see.

History

This spring was once located behind James and Margaret Crawford's cabin. Their cabin was not the first built in the valley. Fur trappers and traders had come through and spent time here and there, leaving their marks in the form of broken-down remains of cabins, similar to the ones that now sit just off the sides of roads and trails throughout Steamboat and the surrounding areas.

The Iron Spring was mentioned in Lulita Crawford Pritchett's *Cabin at Medicine Springs*, in which she recalls her grandparents and their children

Iron Spring with wooden viewing platform. The town of Steamboat was still in the early development stages at this point. Outwest Photo, Boulder. *Courtesy Tread of Pioneers Museum.*

enjoying water from the spring with a twist of lemon. The Soda Spring was also called the Iron Spring for a time.

The temperature of this spring runs approximately seventy-three degrees Fahrenheit, not quite warm enough to bathe in, at least in winter. Bathing was reserved for the sulphur spring across the way from the Crawfords' home.

There is a good possibility the Iron Spring is the same as the Heron Spring, the history of which is listed below.

HERON SPRING, 1900S

Mineralogy

Grains in One Pint Chloride of Sodium	7.59
Carbonate of Iron	2.79
Sulphate of Lime	5.23
Sulphate Magnesia	2.97
Sulphate Soda	2.84
Sulphate Patassa	0.81
Silicic Acid	0.51
Phosphoric Acid	Trace
Manganese Oxide	Trace
Total Solids	22.74

This spring got its name from Mrs. Crawford's pet heron, according to a document housed in the State Historical and Natural History Society:

> *Years ago, Mrs. Crawford, (the wife of one of the founders of the township of Steamboat), came into possession of a young heron. By the time it was grown it was perfectly domesticated. It was his habit, when the morning were cold, to go down early to this spring on a long swinging trot; would plunge the whole of his long, thin legs into the warm water, and enjoy the grateful temperature until the sun was well up. He would accompany his mistress to the church door on Sabbath mornings, and recognized visitors by tapping with his long beak on the front window. Thus it was that "heron" spring came into history. This is also a chalybeate water, sparkling,*

effervescent, emitting a large amount of carbonic acid gas, flowing a larger quantity than any other such spring in the world, superior in every respect to even Pyrmont and St. Morets of Germany, for bath and drinking purposes.

Progress

More than once, what has been considered progress for the town has had negative effects on the natural world. Inevitably, resources are gleaned, depleted or disrupted as towns are built and expansion takes place. When Highway 40 was initially paved, the underground passageways that flowed to the Iron Spring was interrupted, and the Iron Spring became nothing more than a few bubbles working their way up from the earth.

In the event of water shortages from rivers and snowpack, homes that have natural springs on them that provide safe drinking water will be coveted resources. Today, with water coming from taps and many people unaware of the original source of their water, it's easy to take the plenty offered in this country for granted.

Chapter 5

Lake Spring

Location

This spring is the main feature in West Lincoln Park across from the Bud Werner Memorial Library. Turn left from downtown Steamboat on 13th Street. Immediately on your right-hand side is West Lincoln Park. Parking is accessible just below the park. When entering Steamboat from the west, the

The current "Welcome to Steamboat Springs" sign sits adjacent to the Lake Spring and is often dusted with snow in the wintertime. © *Dagny McKinley 2012.*

Lake Spring is the first body of water to greet visitors and welcome locals home. At the edge of the Lake Spring is the official "Welcome to Steamboat Springs" sign nestled under a few pines.

WHAT TO EXPECT

The spring gets its name from its size, which is as large as some of the high alpine lakes that dot the mountains around Steamboat. Because of the warm temperatures, ducks and geese take up residence throughout the year. Dozens of feathered friends can be found floating across the water, ducking down for a meal and filling the air with their boisterous conversation. On the south side of the spring sits a model steamboat, a remembrance of the *Steamboat Spring*, whose siren call sounded like a steamboat to fur trappers unfamiliar with the land. This spring is adjacent to the Sulphur Spring, not to be confused with Black Sulphur Spring. The park is a wide-open grassy expanse. In winter, kids play and build snowmen. In the summer, a wide variety of events take place here, including Art in the Park, theater performances and people playing with their dogs or going for strolls.

Ducks and geese float on the warm waters of the Lake Spring throughout winter. © *Dagny McKinley 2012.*

History

Lake Spring is currently composed of several springs that used to flow through West Lincoln Park, creating a marsh-like area. In order to make better recreational use of the surrounding area, the Sweetwater Spring, or Sweet Spring, was covered and piped into the lake when Highway 40 was expanded in 1988. Other springs were also routed into a bowl of sorts that empty their waters into one location.

According to George Tolles, you can still see where each spring bubbles to the surface at the top of the lake. "There is one spring that surges and bubbles more strongly than the rest, and this is believed to be the Sweet Spring. Some brave souls still try to walk into the lake to test the milky waters. However, they will soon find their legs covered with an oily substance and the white under-layer of the lake bottom, which turns to black sulphur."

In Africa, a lake similar to the Lake Spring erupted with an explosion of sulphur gas, which killed people and animals alike. The records of one of the first explorers of the area noted an area where animals were found dead nearby. Whether this was from the noxious fumes of the Sulphur Cave Spring or a similar explosion at the Lake Spring, we'll never know. As far back as recorded history in Steamboat, the Lake Spring has always been a safe place to explore.

Several smaller springs, including the once-famous Sweet Spring, now make up Lake Spring in the West End Park. © *Dagny McKinley 2012*.

Chapter 6

Lithia Spring or Milk Spring

Mineralogy of Spring, Early 1900s

	Grams in One Pint
Sodium Chloride	18.24
Potassium	.23
Lithium	.38
Sodium Sulphate	9.17
Calcium	10.42
Calcium Carbonate	3.31
Magnesium	7.11
Ferric Oxide	Trace
Silicia	Trace
Total Solids	49.98

Temperature approximately 68 degrees and the flow more than 1,500 gallons per hour.

*Source: "Analyses of a Few of the Mineral Springs"

Mineralogy of Spring, 2012

Temperature	72 degrees
Alkalinity, mg/l	2,242
Bicarbonate, mg/l	2,174
Boron	none
Calcium, dissolved, mg/l	10
Carbonate, mg/l	none
Chloride, mg/l	1,360
Coliforms, -/100 ml (1/100 ml)	none
Fluoride, mg/l (4 mg/l)	3.3
Hardness, mg/l	144
Magnesium, dissolved, mg/l	29
Nitrate, dissolved, mg/l	none
Nitrate/Nitrate, mg/l	none
Nitrite, dissolved, mg/l	none
Nitrogen, mg/l	none
Ph, units	8.4
Potassium, dissolved, mg/l (5,600 mg/l)	125
Sodium absorption ration	84.36
Sodium, dissolved, mg/l (3,000 mg/l)	2,300
Silica, mg/l	none
Sulfate, mg/l (250 mg/l)	589
Arsenic, dissolved, mg/l (0.05 mg/l)	.114
Barium, dissolved, mg/l (1.0 mg/l)	.03
Cadmium, dissolved, mg/ (0.010 mg/l)	none
Iron, dissolved, mg/1 (0.3 mg/l)	none
Lithium, dissolved, mg/l	3.70

mg/l = milligrams per liter
Elements: Maximum recommended by state and federal government in ()
*Source: Tread of Pioneers Museum, "The Springs of Steamboat Springs Walking Tour"

Gossard stonework at Lithia Spring. © *Dagny McKinley 2012.*

LOCATION

The Lithia Spring can be reached by turning south on 13th Street toward the Art Depot off Highway 40. Follow this road past the Black Sulphur Spring and Steamboat Spring. 13th Street becomes 20 Mile Road or CO Road 33. An industrial complex has become the gateway to the spring. On your left you will see a wooden sign marking the entrance to the Lithia Spring Road. The Steamboat Transit Center is just past the turnoff. Don't be alarmed; this is the right place. Follow the road up the hill. Pass through stonework wrought by H.W. Gossard, who had hopes of turning the spring into a worldwide miracle cure for many ails, and park. Aerial photography marks the spring as a white dot within the landscape of Steamboat's industrial sector. Houses sit up the hill from the spring.

WHAT TO EXPECT

My first visit to this spring was before a photography shoot at the Transit Center. When I saw the white waters and read they were sixty-eight degrees, I wanted to climb in to feel their waters against my skin. A strange urge, certainly, but that curiosity still exists. A friend of mine said he had the same urge and did bathe in the waters. He felt an energy greater than that of the Heart Spring at the Old Town Hot Springs. After reading about them, I realized the waters that flow here offer the greatest benefit through being ingested, but I was not brave enough to sample the white waters at that time. As I arrived for my second trip to the springs, I was greeted by close to fifty ravens perched on the wooden fence surrounding the springs. They cawed and flapped their wings, dispersing as I approached and then landing and watching. I felt calm here.

Lulita Pritchett Crawford mentioned in her writings other smaller springs in the area. The ground surrounding the Lithia Spring was spongy and soft, and I often sunk into water up over the ankle of my boot. I found tributaries of the spring and little pools of milky white water. I decided to be brave and try these famous waters. With one hundred beady eyes watching my every move, I dipped a hand into the Milk Spring and brought it to my mouth. I was immediately greeted with the smell of sulphur, which was strong. As I sipped my first sip, I thought of rotten eggs and wondered why or how people could regularly drink from here. I did not know the secret that Steamboat

A family enjoying Lithia water in 1917. *Courtesy Tread of Pioneers Museum.*

local George Tolles has been telling for years. First, one must let the water sit for a day or two to let the sulphur settle. Then, add a twist of lemon or lime to bring out the natural carbonation. The waters were much more palatable after a few days.

I have returned fairly regularly looking for footprints in the snow leading to and from the waters, hoping to run into more locals who believe in the calming effects the Lithia is said to have. As I sit on the stone bench that surrounds the springs, I imagine women in full skirts, dresses and bonnets with umbrellas to protect from the strong rays of the sun, taking a rest and dipping a tin cup into these waters. I imagine a dirt road taking me from one spring to the next and a world that is a little slower than it is today. The spring is well maintained today. Without the power plant for a neighbor, the Lithia Spring looks much like it has looked for years.

History

Lithia is still prescribed by doctors to treat manic-depression. Manic-depression is characterized by mood swings or highs and lows, with the lows leading some to suicide and highs making people feel like they are on top of the world and nothing can stop them until the next crash. Lithia interrupts the swings, helping calm a person down. The amounts of Lithia found in Steamboat's springs are said to improve and balance mood without fear of overdose. An overdose of Lithia can result in fatigue, shakiness, muscle weakness, drowsiness and slurred speech.

Today, Lithia may be known as a treatment for manic-depression and a way to stabilize mood, but there was a time when Lithia was believed to be a cure for just about everything, including gout, rheumatism, arthritis, liver disorders, kidney disorders, ease for inflammation of the bones and joints, skin disease, scrofula, dyspepsia, anaemical diseases and blood disorders. The 1867 British *Pharmacopeoeia* claimed Lithia helped gout, rheumatism and skin diseases. Lithia, at one point, was even tested as a cure for alcoholism and schizophrenia.

Lithia is a non-addictive substance, but side effects can happen at high doses. Fortunately, the amounts found in Steamboat's natural springs are self-regulating. Lithium is the lightest metal in the periodic table and has a silver-white hue that colors the bubbles of the spring, leaves and rocks around the area white. The spring was nicknamed the Milk Spring because of the frothy white color of the waters. The nickname was certainly not for taste.

There are only a few Lithia springs throughout the world. The Saratoga spring in New York was once world-renowned for its waters, yet Steamboat's Lithia Spring has seven times the Lithia content of Saratoga and forty times more than Virginia's famous Buffalo Lithia Water Spring, writes Linda Miller in her article "Lithia Spring," published in the *Steamboat Pilot*. In fact, many have claimed that the Lithia Spring in Steamboat has more of the mineral lithium than any other Lithia spring in the world. Surprisingly, all twelve springs owned and maintained by the City of Steamboat Springs have traces of Lithia. Half of them have the same content of Lithia as our famous Milk Spring. Dr. Newman, a noted German chemist and physician, visited the Lithia Spring early in Steamboat's history and said "it was his judgment that the most learned pharmacist could not compound a more perfect or effective formula for all kidney derangements than the God of Nature pours out to man from his Milk spring."

In the 1930s, the City of Steamboat Springs leased the Lithia Spring, the Old Town Hot Springs (or Bath Spring, as they were once known) and Howelsen Hill to H.W. Gossard. Gosssard had retired from a successful corset business and fell under the spell of Steamboat's springs. He was given permission from the city to bottle and sell the local water. Gossard registered

Lithia Spring was also known as the Milk Spring for the white, milky consistency of the water. *Courtesy Tread of Pioneers Museum.*

Bottling Lithia mineral spring water. It could be Carol Gossard in the picture. H.W. Gossard was working to promote and bottle the Lithia Spring water in the 1930s. *Courtesy Tread of Pioneers Museum.*

The stonework as you enter the Lithia Spring still shows the letter "G" for H.W. Gossard, who once leased the spring from the city and renamed it Miraquelle Water. © *Dagny McKinley 2012.*

Sign to Steamboat Springs' famous mineral waters. *Courtesy Tread of Pioneers Museum.*

the Lithia Spring bottled water as Rocky Mountain Miraquelle water, hoping for national recognition. The bottled water never took off, despite Gossard's passion and belief in them.

Gossard kept the springs under his care throughout World War II and built guard towers at the entrance, so great was their value to him. He even had the trunk of his car lined with stainless steel "so he could carry a good supply of Lithia water to California with him each winter," said George Tolles in an interview with the *Steamboat Pilot*. Today, one can still see the letter "G" engraved in stone at the entranceway to the spring. Gossard gave the spring back to the city in 1935.

By the 1940s, the town celebrated and advertised the healing powers of Steamboat's waters with a sign at each entrance to the city claiming "Steamboat Springs: Natures Great Laboratory." One arrow pointed toward the "Heart Spring & Natural Hot Waters Swimming Pool." The other arrow directed, "Follow the arrows to the Famous Lithia Spring."

Cookie Lockhart, a Steamboat local famous for becoming the first woman inducted into the auctioneers' hall of fame, recalls her childhood connection to the Lithia Spring. She was told that there were only two Lithia springs in the world and the other one was directly below ours. As a child, she believed if you tunneled through the earth you would eventually end up at the world's only other Lithia Spring.

Lithia Spring in the fall. A thirsty dog seeks water from around the spring. © *Dagny McKinley 2012.*

For some, the Lithia waters had a pull that was fatal. In the 1950s, there was a kind of a wild guy who disappeared from Steamboat. "Nobody could find him for a couple of days. People said, 'He always used to go over there [the Lithia Spring],' so they went over and they dredged the spring. In the waters, the man's body was found," remembers Tom Wither. No one knows if he was drunk when he fell in or if he was pushed in or slipped in.

A dead body wasn't the only thing that was found. When the spring was dredged, authorities discovered typewriters from the high school. Kids would steal things from the school, have parties and throw things in the spring. Who knows what lies under the milky waters today?

In 1979, between 50,000 and 100,000 Americans were prescribed or were using lithium as a health remedy.

As with most of the springs in town, there was a time when the Lithia Spring was somewhat forgotten. When the Lithia fad ended, the spring was abandoned without the care of the city. It became littered with trash. The road to the spring was unpaved and left in disrepair. This clandestine spring, out of sight of the rest of town, became a meeting place for lovers.

The year 1987 marked a remembrance of the importance of the Lithia Spring to the city. The Lithia Spring received capital improvements that updated the landscaping and improved signage. The history of this spring is so compelling that to ignore the benefits of it or to lose it behind industrial buildings is to forget the incredible healing powers of the landscape that surrounds us. Each body of water, each plant and each tree in this ecosystem has a purpose. To forget the past is to lose a potential key to our future.

Local George Tolles remembers. He drinks two glasses of Lithia water a day, but only after the water has had a chance to settle. A squeeze of lemon or lime finishes his concoction. Tolles looks good for his years. Between the Lithia water and his soaks in the Old Town Hot Springs, Tolles may just outlive us all.

Other old-timers visit the spring regularly for their water, but not many younger folks do. Hopefully, the next generation will become aware of the healing benefits of these waters, and they will continue to be preserved as they deserve to be.

Chapter 7

Narcissus/Terrace Spring

Mineralogy of Spring, Early 1900s

Analysis	Temperature 73 degrees F
Bicarbonate of Soda	551.65
Bicarbonate of Iron	1.79
Chloride of Magnesium	11.81
Chloride of Potassium	17.89
Chloride of Sodium	201.77
Sulphate of Lime	39.17
Sulphate of Potassium	5.16
Silica	2.68
Total Solids	734.95

Location

Follow directions to the Steamboat Spring or the Black Sulphur Spring. If you are standing at the Steamboat Spring, look for a narrow foot trail heading northeast. The trail sits between the railroad track and the Yampa River. The terrain is slightly rocky but not steep. Take caution, as the rocks can become slick in rain and snow. A slip into the river will cool down a person pretty quickly.

WHAT TO EXPECT

This is one of the few springs in town without any type of signage or viewing platform, which is part of the fun. The springs look practically identical now to when they were photographed in the early 1900s, although the hole worn in the rock by the springs may be a little larger. Glancing out from the springs in a northwest direction, the bridge crossing the river is only slightly different from the one that was first constructed over the water. The bike path is a new addition to the scenery on the opposite side of the river. In this rocky area, there are plenty of perches for sitting, taking in the views, people-watching and relaxing to the sound of water flowing by. In the winter, these springs are easy to locate from the library side of the river. The Terrace Spring is the larger flow of water that melts out the snow in winter and makes the rocks shimmer in spring and summer. Steam gives away its location as it rises upward and outward from deep below the surface of the earth.

The Bitter Spring/Terrace Spring. *Courtesy Tread of Pioneers Museum.*

HISTORY

Both the Narcissus and Terrace Springs exit out of rock formations where they flow into the Yampa River. The mud that surrounds these springs was once used to treat skin disorders such as eczema and other skin conditions and irritations.

A current view of the Terrace and Narcissus Springs. © *Dagny McKinley 2012.*

The Narcissus Spring is the one between the Steamboat and Terrace Springs and is closer to the river than the Terrace Spring. From the Narcissus Spring flows a clear, sparkling stream of water, "iridescent with blue-green, rose and silver minerals deposited in the stream and quietly rolling, deep fairies' pool," according to the *Steamboat Pilot* article "Great Variety of Mineral Waters Have Many Tastes," July 30, 1957. Much like the mythical Narcissus, one who stares too deeply into these waters may end up falling in love, either with the clear reflection seen from them or with the waters themselves, and be fated to waste away here.

The Terrace Spring was once referred to as the Bitter Spring due to the high quantity of sulphur and alkalis contained in it. One of the first records of the spring, courtesy of the Steamboat Springs High School Library, reads:

> *A Bitter, pungent taste, more pronounced in this than any other of this group, has fixed its name resembling somewhat the great "Bubbler" in its chemical aspect. The increased "bitter" being due to the predominance of the powerful alkalies. Attention is called to the fact that the solid matter of this spring is greater than any other. From a miniature cavern the crystal rivulet, together, with the odor of various gases, comes forth into the outer world, making in its brief flight a splashing little cataract as it plunges over a low precipice into the Yampa.*

The spring tumbles over terraces into the Yampa River. One of the earliest brochures on the subject, "Mineral Springs Have Great Potential Value," stated the Terrace Spring was believed to contain the highest percentage of soluble sulphur of any spring in Colorado. Once believed to cure blood and skin diseases, this spring has a place in history, the brochure went on, and then finished by noting, "A Scotch chemist declared that this water 'contained the most perfect chemical solution of sulphur he had ever tested.'"

Chapter 8

Old Town Hot Springs

Mineralogy of Springs, Early 1900s

Bath Spring

	Grains in One Pint
Carbonate of Lime	1.71
Sulphate Soda	0.69
Sulphate Magnesia	0.76
Sulphuric Acid	0.07
Silicic Acid	2.79
Phosphoric Acid	Trace
Potassa	Trace
Alumina	Trace

Temperature: 103 degrees F
Flow: 225 Gallons per Minute
*Source: "Analysis of a Few of the Mineral Springs"

Mineralogy of Springs, 2012

Temperature	102 degrees
Alkalinity, mg/l	183
Bicarbonate, mg/l	103
Boron	.7
Calcium, dissolved, mg/l	18
Carbonate, mg/l	84
Chloride, mg/l	320
Coliforms, -/100 ml (1/100 ml)	none
Fluoride, mg/l (4 mg/l)	1.9
Hardness, mg/l	49
Magnesium, dissolved, mg/l	1
Nitrate, dissolved, mg/l	none
Nitrate/Nitrate, mg/l	none
Nitrite, dissolved, mg/l	none
Nitrogen, mg/l	none
Ph, units	8.0
Potassium, dissolved, mg/l (5,600 mg/l)	11
Sodium absorption ration	n/a
Sodium, dissolved, mg/l (3,000 mg/l)	300
Silica, mg/l	49
Sulfate, mg/l (250 mg/l)	150
Arsenic, dissolved, mg/l (0.05 mg/l)	.005
Barium, dissolved, mg/l (1.0 mg/l)	none
Cadmium, dissolved, mg/ (0.010 mg/l)	none
Iron, dissolved, mg/1 (0.3 mg/l)	.04
Lithium, dissolved, mg/l	.35

mg/l = milligrams per liter

Elements: Maximum recommended by state and federal government in ()

*Source: Tread of Pioneers Museum, "The Springs of Steamboat Springs Walking Tour"

Location

The Old Town Hot Springs sits on the opposite side of town from the majority of the other springs. They are located on Highway 40 as you enter town from the south and are on the eastern side, just before you reach the post office at the corner of 3rd and Lincoln.

What to Expect

The Old Town Hot Springs have always been the center of community activity. From the time James Crawford discovered the springs, an object of much appreciation for warm baths in a frontier town, until now, people have used them as a place of congregation and relaxation. The hot springs have become the place to go to train for swim lessons, kayaking lessons, first aid courses and, every now and then, as a place for a dog to take a dip. The Old Town Hot Springs are owned by the SSHRA, whose mission is to keep the springs available to the community. As a result of the SSHRA's ownership, the bathing and swimming pools have been expanded into a facility that offers massage, fitness lessons, a workout room, a daycare center, kiddie pools, slides and even a private hiking trail. This is where locals run into one another, families spend quality time and visitors come after a long day on the slopes to relax in the famous Heart Spring. There is an entrance fee to use the springs. Passes can be bought annually or daily. Over fifty thousand people visit these springs each year.

History

The Old Town Hot Springs have undergone several name changes during their history. The original name for these springs was the Bath Spring, which held until the 1930s, when the spring was renamed Heart Spring for marketing purposes. At that time, a heart-shaped rock wall was built around it. As the city expanded closer to the ski mountain, the spring became known by its current name, Old Town Hot Springs.

The Ute Indians, who believed the springs had healing powers, originally used the Bath Spring. The Utes only summered in the Steamboat Valley, heading south when the snows came. Because of the springs and the vast hunting opportunities in the valley, the land was coveted by other American Indian tribes, and battles occasionally took place. According to legend, the famous Chief Yarmonite, head of the Yampa Utes, was a child when a war party of Arapahos surprised the Utes and attacked. The Utes were camped on an island on the Bear River (Yampa River) close to the Bath Spring when the enemy attacked. Many Utes died that day, including Chief Yarmonite's father. Chief Yarmonite later recounted to James Crawford that he felt he must return to the valley every summer to honor his father. This story may be the basis for the Yampa Valley Curse, which states that once you visit the Yampa Valley you will never be able to leave, and if you do, you will return. Locals, myself included, swear they are pulled back time and again.

An early description of the Bath Spring from the collection of the Steamboat Springs High School Library paints a picture of the early days of the springs:

> *BATH SPRING*
> *The name and temperature of this spring indicates external use. Whoever is fortunate enough to bathe here once, longs to return, and is tempted to linger in the luxurious pool beyond the prescribed time. No Roman even in the days of the opulent splendor of the Caesar's ever enjoyed a luxury equal to this. The discharge is about 225 gallons per minute, equal to two-thirds of the total amount discharged by the entire group at Hot Springs, Arkansas.*

When the town was first settled, this spring was not easily found. According to Lulita Crawford's account in *Crawford Pioneer Tales*, the discovery went as follows:

> *Of all the springs, perhaps the bath spring was the one most enjoyed. This spring was not discovered for more than a year after Pa had preempted his cabin. The family was still living at hot sulphur spring (located between Kremmling and Granby, CO), but in late summer 1875 made a camping trip to Bear (Yampa) river to see where their new home would be. While Pa was hunting deer on the sagebrush slopes at the foot of Storm Mountain on Aug 1, he followed a game trail down a hill to a small creek and blundered into a bubbling spring in the middle of the stream. The water was hot but not hot enough to scald his horse. He thought it would be just right for bathing. Pa went back and brought the family. Kids bathed first, then women and then men.*
>
> *The only other spring that had heretofore been used for bathing was the bubbling sulphur spring at the opposite end of the valley. When Pa had*

Steamboat Bathhouse, the first bathhouse built over the Old Town Hot Springs. *Courtesy Tread of Pioneers Museum.*

> *discovered that spring the previous summer it had already been hollowed out and rocks laid around it, an indication that Indians or mountain men had used it. It was only tepid.*
>
> *Pa turned the channel of the creek away from his new find and built a small log house over it.*

Thus, the spring became known as the bathhouse. As the population expanded, the spring was developed with indoor and outdoor pools and other forms of recreation. Originally, Spring Creek ran through the Bath Spring, but it was later diverted to the west of the spring.

When Lulie Margie Crawford, daughter of James and Margaret Crawford, headed to the spring in 1881 for a soak with her uncle, they found the housing over the bathhouse caved in. Her Uncle Henry found a stick and propped up the roof so they could still bathe. Guests enjoyed bathing in the spring as much as the Crawfords did. The diary of Lulie Margie Crawford notes,

"There was a standing joke that the first thing the Crawfords always invited any guests to do was to take a bath. This invitation was readily accepted since to soak in a big, bubbling spring of comfortably hot mineral water was a rare treat in a country where the ordinary bath was a hurried scrub in the washtub in the kitchen."

When the Crawfords first moved to town, grizzly bears populated the area. Logan Crawford told the tale of running into five grizzlies one morning on his way to take a bath in the hot spring at the other end of town from their home. His first encounter was with two big male silvertips, as recorded in *Great Good Old Days* by John Daniel Crawford: "Logan ducked and when they passed he ran in a crouching run to detour over the hill where he almost bumped into a mother grizzly and her two cubs. He lay in the brush and stayed still until they finally passed." That day, he came to the conclusion that Bear River was aptly named.

Improvements gradually were made to the spring to make it more hospitable. A sluice gate was used to control Spring Creek before it was diverted into the river. Bathers hung their clothes on nearby trees while they bathed. Always the visionary, James Crawford saw the spring as bathing and swimming pools that should be developed for recreation. Crawford recruited A.J. Macky, James P. Maxwell, Lewis Cheney, Harry A. Burgess and others from the town of Boulder to form the Steamboat Springs Company. They constructed a frame building over the spring. In 1884, the bath spring had its first employee. A caretaker was on site to hand out fresh towels and soap for bathers.

In the very first issue of the *Pilot*, Steamboat's first newspaper, an announcement appeared:

> *A new bathhouse, with splendid plunge bath, dressing rooms, etc. has just been completed at the Hot Spring, and until further notice is open for the free use of all who may come. The pioneers of Bear River and all citizens of Routt County are especially invited to come early and often and move the waters of this Bethesda and go forth cleansed and rejoicing.—J.H. Crawford, for Steamboat Springs Company*

A key hung on a nail outside. The bather took the key, and when he left, he put it back on the nail. People came from as far away as Hahn's Peak, twenty-five miles north, to use the spring.

At the time, while bathing was free, other goods ran a much higher price. According to a grocery list that was preserved from the time and reprinted in the *Pilot*, "a quart of syrup cost $1.25; coffee, 1 lbs., $1.60; pork, 4½ lbs., $4.15; sugar, 2 lbs., $1.35; sack of flour, $26. With luck, $10 could get a homesteader through the winter as long as they had wild game to supplement their supplies."

In 1887, logger Horace Suttle, who owned a sawmill nearby, constructed a newer frame over the bathhouse. It was simple and had one room but was a change for the better. Two years later, three additional rooms were added.

In 1909, ownership of the bathhouse was sold by the Steamboat Springs Company to the Steamboat Springs Town and Quarry Company, a group of men from Nebraska. They began construction on the bathhouse, which was big news. The *Pilot* reported on the Albert Ohamn force of men making "rapid progress." According to the article, a rock crusher and cement mixer were shipped to town by the Town and Quarry Company. The goal was to have the renovations completed by July 1 in time for the national holiday. What a different July 4 this would be compared to the one first celebrated by the Crawfords with the Ute Indians during their first year in Steamboat.

An investment of $50,000 was made to transform the spring into a recreation area. A new eighty-four- by one-hundred-square-foot bathhouse was built of stone from the surrounding area, as was an outdoor pool. The minerals from the springs worked away at cement and wood. During the first year, revenue was expected to only be about $187. As Lulita Crawford observed, "It seemed as though the waters fought against being tamed."

The *Pilot* article "Colorful History of Local Swimming Pool Had Its Start Back in 1884" reprinted an account of one of the first years after the train came to town:

> *The year 1909 brought many good things for Steamboat Springs, giving this town an impetus and growth which will forever clinch its position as the metropolis and commercial center of Northwestern Colorado. The railroad came early in the year, placing us within a few hours of the center of the commercial life of the state. The long-needed hotel was built. The bath house and pool, splendid pieces of work in themselves, are only at the beginning of the work of improving the springs and making them useful and attractive to man. Many fine business blocks and scores of attractive residences were completed, the total building record showing close to a quarter of a million dollars. And not the least in importance to the town was the development of the strawberry industry to a paying basis and the certainty of its wonderful expansion in the next few years.*

Growing strawberries was a lucrative business and for a while big news. Les Remington is reported to have had a six-acre patch of strawberries that he sold to Denver for a $3,000 profit. The late growing season at high altitude gave Steamboat an advantage in the market because its berries grew later in the season than other berries did. Steamboat's strawberry business flourished, and the berries were shipped as far as Chicago. For three years the business boomed, but those years were followed by three years of an early frost that effectively put growers out of business.

As the town grew up, upgrades were made to the Bath Spring, and the first fee was charged. Bathers paid ten cents for admission, but the rules weren't too strict. Kids were given the option of pulling dandelions for a few hours in exchange for admission.

An unknown source reported:

> *The first bath house was unique. Mr. Baird was the manager. The building was long. One entered an office with rooms in the back for dressing. Back of these was the "bathing room." People actually took baths there. Mr. Baird assigned you to a dressing room, which you could lock so no one could enter while you were using the pool. You then returned to your dressing room, locked your door and rang a bell. When Mr. Baird heard the bell he knew you were through and would let the person in the next dressing room out to bathe.*

Marcellus Samuel Merrill, who grew up in Steamboat, used to visit the Old Town Hot Springs with his parents once a week. They paid fifty cents to go in a room and scrub down, as he recalls in *Steamboat Springs: Memories of a Young Colorado Pioneer*.

In 1910, the bathhouse was turned into an indoor swimming pool. At some point, due to financial troubles, the pools were deeded back to the Steamboat Springs Company.

Humans weren't the only ones to explore the Old Town Hot Springs. In February 1917, an elk took a dip in the pools, to the delight of the townspeople. The creature's misadventure was recounted by the *Pilot*:

> *Wild Elk Enjoys Plunge in Local Artificial Lake*
> *Animal separated from fellows visits the town in broad daylight, swims in pool at Bath House and is finally added to the herd at local park.*
>
> *A portion of Steamboat Springs' population was privileged to witness the novel sight of a wild elk taking a bath in a public bathing pool within city*

Indoor pool at the Old Town Hot Springs, circa 1920. *Courtesy Steamboat Springs Health and Recreation Association.*

limits. This opportunity was furnished Saturday when a stray bull elk wandered into town.

The animal came from the north and doubtless was a member of a small band that has been seen several times of late in the vicinity of Hot spring. He was seen passing the VaBrunt place, then about 11 o'clock near the Light residence, from which point his general course was down Spring creek to Lincoln avenue. Here a couple of dogs gave chase upon perceiving an animal so much out of the ordinary range of their observations. Anxious to avoid attack the elk galloped up the avenue toward the bath house. Near this point one of the dogs by a flank movement swerved the wild creature's course and for safety he ran onto the big bath house porch, where he stood frightened, panting and uncertain. The dogs evidently considering that they had done their full duty in the premises by driving their victim to cover stood guard but did not offer pursuit.

About this time some patrons of the bath house opened the front door to come onto the porch. Startled, the elk looked first at the new danger that threatened, then back at that from which he had so recently escaped. Something must be done, and that quickly. Hemmed in before and behind, but one course remained, and after a second's hesitation he took it. Rising gracefully in the

An elk takes a dip in the Old Town Hot Springs. *Courtesy Steamboat Springs Health and Recreation Association.*

air the animal cleared the railing and landed in the outside pool, where he gave an exhibition of swimming such as had never been seen there before. Finally, seeing that no one pursued him, he settled down in the shallower portion of the pool, stood still and scrutinized the strange surroundings.

Men from the forest service were notified and shortly after noon took charge of the unusual proceeding. After a few ineffectual attempts the elk was roped and brought to terra firma. A transfer sled was utilized to transport him to Elk park, where he was cautiously set at liberty to join other members of his tribe that inhabit the place as guests of the town of Steamboat Springs. For several hours the animal showed no disposition to hunt company, but stood at the gate, sadly gazing toward the scene of his unusual escapade. Brothers and sisters who have made Elk park their home for two years came down, had their evening meal and went back into the hills, but the stranger stood by the gate. Sunday morning he was still on guard, but later disappeared, so it is probable he has now joined the band, accepting the inevitable.

Forest service men have been scouting the hills for several days in search of other elk and deer which have been seen in the neighborhood during the last few weeks, who beside the one capture were all they were able to account for, so there is a suspicion that some one has been violating the game law. The fine big fellow that came for a bath was frightened by hunters, the

Today's Iron Spring is in a marshy area just off Lincoln Avenue, the main road through town. © *Dagny McKinley 2012.*

The Lithia Spring was also known as the Milk Spring for its white, milky color. © *Dagny McKinley 2012.*

The high content of calcite in the Lithia Spring turns surrounding rocks and leaves white. © *Dagny McKinley 2012.*

Carbonated waters that may be from the Soda Spring trickle onto the banks of the Yampa River. The effervescent waters have a sparkling, refreshing taste. © *Dagny McKinley 2012.*

Left: Winter snows blanket the area surrounding the Old Town Hot Springs for six to eight months of the year. © *CoreyKopischke.com.*

Below: The Old Town Hot Springs is the first sight to greet visitors traveling north through town. © *CoreyKopischke.com.*

Left: Poochy Paddle 2010–11. Dogs are welcome at the Old Town Hot Springs on the last day of the ski season before the pools are cleaned. *Right*: The Old Town Hot Springs' version of the Cardboard Classic challenges kids to build boats with cardboard and duct tape and successfully float them across the pool without them disintegrating. *Courtesy Steamboat Springs Health and Recreation Association.*

The Steamboat welcome sign commemorates the famous spring that could once be heard for miles. © *Dagny McKinley 2012.*

Left: Next to the railroad that once brought visitors and commerce to Steamboat, a small spring bubbles. © *Dagny McKinley 2012.*

Below: As the railroad from Denver heading west was being constructed, builders threatened to bypass Steamboat until the townspeople raised money to buy themselves a stop on the line. That decision cost Steamboat the only geyser in Colorado. © *Dagny McKinley 2012.*

The springs that now feed the Strawberry Park Hot Springs can be seen steaming out of the hillside. © *Dagny McKinley 2012.*

Cabins at the Strawberry Park Hot Springs can be rented by the night. Lodgings are rustic, but guests can soak after the springs close to the public for the night. © *Dagny McKinley 2012.*

This rusted car welcomes guests to Steamboat's famous hot springs. © *Dagny McKinley 2012.*

In the fall, there are fewer crowds at Strawberry Park Hot Springs than summer and winter, when the pools are filled with families and friends. © *Dagny McKinley 2012.*

Strawberry Park Hot Springs has three pools of varying temperatures. In the lower pool, guests can cool off by jumping in a mix of river water and hot springs water. © *Dagny McKinley 2012.*

A current view of the Strawberry Park Hot Springs. © *Dagny McKinley 2012.*

Bathers can soak in one of several private pools at Strawberry Park Hot Springs. © *Dagny McKinley 2012.*

This teepee that serves as a changing room for guests to Strawberry Park Hot Springs is one of the few reminders of the Ute Indians who used to call the Yampa Valley home. © *Dagny McKinley 2012.*

An old cabin reminds hikers along the Hot Springs Hiking Trail of the pioneers who were the first white people to settle the Yampa Valley. Most cabins are now long overgrown. © *Dagny McKinley 2012.*

A tree on the hiking trail to the Strawberry Park Hot Springs is decorated with animal skulls. © *Dagny McKinley 2012.*

The Sulphur Cave holds many mysteries not yet known to science. Snottites were recently discovered and are only known to exist in one other cave in Mexico. © *Dagny McKinley 2012.*

The Sulphur Spring, also known as the Bubbling Spring, still captivates kids and adults alike. © *Dagny McKinley 2012.*

Above: The mud from the Narcissus/Terrace Springs was used on skin to heal skin disorders and create clear skin. © *Dagny McKinley 2012*.

Left: The Terrace Spring was once called the Bitter Spring for the high content of sulphur and alkalis. © *Dagny McKinley 2012*.

Opposite top: The Terrace Spring was believed to cure blood and skin diseases. According to a Scotch chemist, this water "contained the most perfect chemical solution of sulphur he had ever tested." © *Dagny McKinley 2012*.

Opposite bottom: Views from the Terrace Spring haven't changed much from the early 1900s as the waters flows into the Yampa River. © *Dagny McKinley 2012*.

Left: Water flows from a spout to heat the Strawberry Park Hot Springs' various soaking pools. © *Dagny McKinley 2012*.

Below: The Fetcher family has a long legacy in Steamboat Springs. In this picture, Bill Fetcher's mother, Criss (Clarissa) Fetcher (1915–2005), enjoys a picnic at the Strawberry Park Hot Springs in 1961. © *Henry C. Wells (1876–December 1967)*.

Top: A swimmer at the Old Town Hot Springs. The Olympic-size pool is heated by the Heart Spring. In winter, the pool closes early due to subzero temperatures. © *Dagny McKinley 2012.*

Left: Lily Frinks climbs the rock wall at the Old Town Hot Springs. © *Dagny McKinley 2012.*

Left: As part of an art installation representing the walls we build for ourselves and how love can tear them down, a cocoon was hung in a sheltering tree along the hiking trail to the Strawberry Park Hot Springs. © *Kim Keith Photography 2012.*

Below: The slides that are situated high above the fence of the Old Town Hot Springs were once a topic of debate in town. Some believed the slides created too much of a Disneyland feel at the entrance of town. © *Dagny McKinley 2012.*

When H.W. Gossard leased the Old Town Hot Springs from the City of Steamboat Springs, he built a heart-shaped pool around the main spring and named it the Heart Spring, as seen in this photo, circa 1920. *Courtesy Steamboat Springs Health and Recreation Association.*

> *officials believe and it may be that a few persons not far from Steamboat are now feasting on elk meat. But Steamboat has another fine specimen for its park as the state doubtless will grant authority for the city to retain possession of the elk secured in this unusual manner.*

In 1929, the bathhouse was given to the University of Colorado until 1931, when H.W. Gossard leased the Old Town Hot Springs, Lithia Spring and Howelsen Hill from the city. He renovated the spring with marketing in mind. Around the source of the Bath Spring, he built a heart-shaped wall made of river rock and named the spring the Heart Spring. He also added a second floor where lockers and storage were provided.

Gossard tried his best to make this area a tourist destination through brochures and a Water Carnival. The very first carnival took place in September 1931 and attracted famous swimmers and divers from all over the United States. Despite Gossard's attempts at creating a spa area for people from around the country, locals remained his most faithful patrons. At one point, a skating rink was added to the facility, but with little success. There was, however, one show that always brought in the crowds. Former carny (carnival worker) Ray Woods would dive off a one-hundred-foot tower into the pool in order to bring in business. Wood and his wife ran a concession stand on the premises.

Woods wasn't the only one doing daring dives. According to Chuck Leckenby, a longtime Steamboat resident whose family owned the *Pilot* newspaper for over one hundred years, there used to be three diving boards. There were two one-meter boards and one three-meter board. Above the three-meter board there was a ladder that went up to five meters above the pool. Olympic divers dive off the five-meter tower, or fifteen feet above the water. The more daring kids used to get on the tower, which was only about two by three feet on each side, and then jump on the three-meter board and dive into the pool.

Gossard's legacy is a community center that serves the local families and visitors to the valley. He created a fun, family atmosphere that is continued to this day. Pat Carney, executive director of the Steamboat Springs Health and Recreation Association, the current nonprofit stewards of the hot spring, says:

> *Our location and variety of amenities make us a community gathering place where all ages come to have fun and get healthy. We are not just a hot springs; the fitness center and the lap and kiddie pools attract a wide variety of people. The schools have programming here, as well as the day camps, after-school programs, pre-schools, etc. We are stewards of the hot water but have used that to create so much more.*

In the 1940s, Chuck Leckenby and a few of his friends used to climb up on the roof of the old facility at night, drop into the indoor pool and then sneak out. Other fun was had as well. Freestyle skiing may well have been invented in Steamboat as some of the ski jumpers used to fly down the hill behind the spring on their skis and ski across the water.

In 1935, after just a few years and due to the Depression, Gossard returned the hot spring to the Steamboat Springs Town Company, the original owner. The *Pilot* article "Colorful History of Local Swimming Pool Had Its Start Back in 1884" reports:

> *Not only was the swimming pool and its adjacent land concerned in these dealings, but also the land now in the rodeo grounds, the ski hill, the land where the golf course and Lithia spring are located and the Hot spring seven miles out of town. Many things had happened during use of this property for various purposes and a group of civic-minded men had the idea that it might be wise if the town could some way have control of the property. After a grandstand, which had been erected by the town at*

Old Town Hot Springs, circa 1914. Diving boards were once in place but were removed due to safety concerns. They may be reinstated in the future. *Courtesy Steamboat Springs Health and Recreation Association.*

Skeeter Werner and a skier, circa 1950s. Pictured are the large outdoor pool and the present lap pool at the Old Town Hot Springs. *Courtesy Tread of Pioneers Museum.*

> *the rodeo grounds was ordered dismantled and sold at public auction, the idea crystallized.*
>
> *C.R. Monson and Claude Lukens were the first to approach the Boulder men with the idea and in dealing with Ralph Newcomer and University of Colorado reached an agreement whereby the property would be sold to a non-profit organization for $25,000, the amount of indebtedness and taxes that were outstanding.*
>
> *Accordingly the Steamboat Springs Health and Recreational Association was formed as a non-profit corporation to purchase the property. The rodeo grounds, ski hill and golf course were deeded directly to the town of Steamboat Springs and the swimming pool property and Hot spring kept with the association.*

The deal between the town company and the SSHRA was known by some as one of the greatest bargains to fall into the lap of the community. Roy P. Hofstetter, previous secretary of the SSHRA, recounted the story for the *Pilot* in 1952:

> *The heirs of the Maxwell and Mackey families sold the swimming pool, 40 acres of land surrounding the Hot spring, 270 acres of ground that make up the rodeo grounds, the ski hill, two parks and a large part of the golf course as well as 462 town lots, for $25,000.*
>
> *The Health and Recreation association was organized in 1935 to administer the property for the benefit of the community and since that time has paid…the indebtedness as well as spending many thousands of dollars for improvements.*
>
> *All of the property, including the numerous springs, was given to the town, but the Health & recreation association assumed control of the plunges and mineral water pools and the Hot springs property.*
>
> *The original board of the association formed to conduct the great civil asset for the benefit of the community was comprised of C.R. Monson, Claude A. Luekens, Dr. E.A. Jackson, T.W. Poulson, Roy P. Hofstetter, who are still alive and active in the association and late John M. Childress, Charles Baer and A.W. Heyer. Mr. Hofstetter pointed to the membership rates which give children benefit of all the public pools for $4.00 a year and adults $5.00 for the season.*

By 1968, the roof had collapsed over the indoor pool, and it was decided to remove the indoor pool for good, as well as the individual hot

Locals line the edges of the Old Town Hot Springs pool, circa 1947. *Courtesy Steamboat Springs Health and Recreation Association.*

baths. A sauna was added to the structure, and a large sign at the entrance proclaiming "Swim/Sauna" became a local landmark. Approximately $200,000 was spent to remodel the outdoor pool, add regulation lanes for races and add a filtering system so the water could be recycled instead of subverted to the river. An area was built for younger kids, and a new building was added for lockers. The SSHRA looked to the community to support the project and help it receive federal financing. The townspeople stepped up by buying two hundred memberships. Today, more than four thousand locals own memberships.

Pat Carney arrived as manager of the Old Town Hot Springs in 1975, when it was a much smaller place. She recalls, "At that time it was a covered rectangular hot pool and the lap pool was only available for use in the summer." Today, Carney is the executive director of SSHRA and in many ways a guardian of the hot spring. Her direction has been responsible for the current look and feel of the facility.

In 1981, as the Strawberry Park Hot Springs was falling into disrepair with people living for months along the banks and leaving mattresses and garbage around, the SSHRA made the decision to sell the Strawberry Park

Hot Springs to Chicagoan Don Johnson. The money from the sale was used for improvements at the Old Town Hot Springs.

Steamboat residents who had known the spring became involved in the association to help preserve them. One of those locals was Dorothy Wither, who was honored for her work on the board of the SSHRA in 1983 with a commemorative park. She swam in the hot spring waters every morning until the day she died and was responsible for getting George Tolles interested in Lithia water. He would fill up jugs for her regularly. Wither should also be noted for establishing the Tread of Pioneers Museum, as well as teaching all her nieces and nephews to swim in the Old Town Hot Springs. Her nephew Tom Wither worked for a summer cleaning out the private pools that were about six by eight feet, where couples and families could soak in peace, although it wasn't until Tom was an adult that he snuck into the hot spring facility for a private swim—without getting caught.

One of the biggest improvements made with the reconstruction was that bathers were once again able to bathe directly in the Heart Spring, which had been off limits for eighty years. "I was influenced by the late Dorothy Wither," said mineral springs guardian George Tolles at the inauguration of the park. "We all used to have our own key, and we'd arrive at the pool at 5:30 a.m. and let ourselves in. We'd lock up on the way out. Wither was known as Steamboat's 'first lady.' She enjoyed the soothing waters of the mineral spring [Heart Spring] from the time the first bathhouse was built in 1909 until her death at age eighty-four," recounted Tolle. He continued:

> *Dorothy's father, Archie Wither, a native of Scotland, arrived in Steamboat Springs on a freight wagon on the Fourth of July 1889. He homesteaded Emerald Mountain across the Yampa River behind the present-day ninety-meter ski jump on Howelson Hill.*
>
> *A true pioneer, Dorothy embodied the character and strength of the settlers of the Yampa Valley. She loved and enjoyed its natural beauty and above all the waters of Heart Spring.*

Tolles himself has enjoyed the rejuvenating powers of the Old Town Hot Springs since the late 1950s, when he used a series of private wood baths. No clothes were needed, and a wooden sluice gate changed out the water from one bather to the next.

Another longtime Steamboat resident, Anabeth Lockhart, remembers when admission was only twenty-five cents and there were no lifeguards. "The water wasn't always clean and sometimes logs would float in the pool,"

With the Heart Spring restored thanks to the Steamboat Springs Health and Recreation Association, bathers are now able to soak directly in the Heart Spring, which was off-limits for eighty years. © *Dagny McKinley 2012.*

Lockhart told Tom Ross for his article "Lithia Water Mellows the Man" printed in the *Steamboat Pilot & Today*.

Old Town Hot Springs became the official name of the springs in 2006.

In 2007, a $5 million improvement plan began that included a slide for kids, renovated pools and more. The pools were shut down from April to

West bathhouse, circa 1914. The views of downtown show the growth of town and the preservation of these waters. *Courtesy Steamboat Springs Health and Recreation Association.*

Today's view of downtown from above the outdoor swimming pool. © *Dagny McKinley 2012.*

November so they could get the facility opened for the winter ski season. New water slides were added, and the estimated cost rose to $8.5 million. Landscaping and natural elements were added that gave the facility a more natural feel, akin to what it may have once looked like. The improvements keep the tourists visiting, which in turn helps keep costs down for the rest of the community.

Locals weren't sure they wanted to embrace an expansion or even a temporary closure, but as Pat Carney said in an interview with the *Steamboat Pilot*, "We look at it as a resource that we are the guardians of, and we are trying to manage it responsibly and make it available to as many people as possible."

People were concerned with how renovations would affect the Heart Spring. Carney said in the *Pilot*, "We are not touching the Heart Spring. We are moving a few of the edges of it around and tying everything in around it. You couldn't get in there because it was too deep and there were no benches, and it was too dangerous…When we redid it, we created this false floor of flagstone so water could bubble up in between." In a personal interview, Carney expressed the importance of the Heart Spring to Steamboat's history and to the Old Town Hot Springs' future: "We

Water runoff from the Old Town Hot Springs flows into the Yampa River, where people floating the river in inner tubes in the summer often stop to warm up. © *Dagny McKinley 2012.*

A 2012 view of Steamboat Springs from Mount Werner. © *Dagny McKinley 2012.*

honor the Heart Spring in everything we do—it is the reason we're here and the one of the main reasons we continue to be successful. It will always be the centerpiece."

During construction, new hot springs were found, as well as a one-hundred-year-old clay water pipe. The pipe was discovered when it was hit during construction. Workers also found a leak in the Heart Spring pool that was pumping out up to one hundred gallons per minute. The Heart Spring is the source that heats all the swimming and soaking pools at the facility. Any runoff is piped under the highway to a pool formed by rocks in the Yampa Valley. In the summer, when tubers float down the Yampa, they often stop for a warm-up soak in the springs' runoff.

Part of the new renovations includes a rock-climbing wall that edges the end of one of the pools. Kids can climb without fear of falling into anything other than water. A waterfall of sorts flows down the wall, making climbing a little more challenging. And then there are the huge slides that have become part of the landscape of driving into town.

As with any area heavily used by the community, everyone has his or her own opinion over what the place should look like and how it should be represented and used by the community. In a letter to the *Pilot* in April 2007, David Scully takes the planning commission to task:

Sunday, April 1, 2007

In response to the article last Friday that discussed the Planning Commission's rejection of the proposed waterslides in the Steamboat Springs Health and Recreation Associations' pool renovations, I would like to offer the following observations:

Commissioner Lewis says that it is "too much Disneyland" at the entrance of downtown to approve the tower and waterslides. Currently there is a big pink sign with a rabbit on it across the street to "the entrance of downtown." Is the Rabbit Ears Hotel sign exemplifying any image of Western? Lewis states it would not be good for retail stores F.M. Light and Allen's, stores that apparently exemplify Steamboat's ranching heritage according to the article.

How do you know that these businesses would not welcome additional sales that they may receive by visitors who are discovering that there is a hot spring and recreation pool facility, because of the visible tower and waterslide addition to the SSHRA, which also has led them to discover that Steamboat actually has a downtown with stores like F.M. Light and Allen's.

As for the comments of SSHRA losing its natural feel, there is always Strawberry Park Hot Springs up the road for those wishing to commune with nature while they soak sans waterslide, kiddy pool, lap pool and noise from Lincoln Avenue.

Walt Disney created an entire theme to recognize our country's history from which pioneering led to manifest destiny and the creation of what is perceived as "Western heritage." He called it Frontierland and it was part of the original Disney theme park in Florida, Disneyworld. Frontierland was made to re-live the pioneer days of the American frontier. Walt said, "All of us have a cause to be proud of our country's history shaped by the pioneering spirit of our forefathers…our adventures are designed to give you the feeling of having lived, even for a short while, during our country's pioneer days."

It seems Walt knew something about creating fun around our heritage and history. He certainly understood what it meant for a business to create itself as a destination that people would travel to for entertainment. This is a theme that the Steamboat Springs Mainstreet organization has promoted to help the downtown business community create themselves as destination businesses that visitors and locals will recognize and be drawn toward. Walt had to build his Splash Mountain.

Howelsen Hill added its slide attraction several years ago. It seems to be a successful destination, even though it really doesn't blend in with the natural vegetation on the hillside.

The waterslides are not about interpreting what is Western heritage, or how to preserve it. It wasn't a threat to Steamboat's Western heritage when the ice rink expanded. Steamboat is not in danger of losing its Western label because Strings in the Mountains is building a permanent structure. These examples demonstrate how Steamboat has grown to ensure amenities that are part of Steamboat's evolving heritage are perpetuated. Steamboat has a history of ranching. Geographically, Steamboat is in the West. These are not facts that will be lessened by waterslides at the "entrance to downtown."

It's not about a Western heritage; it's about supporting businesses and amenities that are part of Steamboat. The springs and hot pools that SSHRA manages have been in place for more than 100 years. SSHRA needs to evolve again, to adapt and grow as Steamboat does. With the multitude of developments that will change how downtown looks, this is literally a landmark opportunity for SSHRA to showcase what it has to offer.

The SSHRA board of directors appears to have its pioneering hats on in the spirit of mapping the future. Lewis and Curtis appear to have lost their way.

David Scully
Steamboat Springs
Kayaking courses—lifeguarding courses

The slides went in, and the grounds look beautiful. More importantly, they are enjoyed. For the grand opening, admission was free to the public for four hours to enjoy the new amenities, including a new 3,000-square-foot hot pool; an aquatic climbing wall; a pair of 230-foot water slides; a 240-square-foot spa; a refurbished, 103-degree Heart Spring pool; a new glass-enclosed slide tower; a remodeled lobby; and a new cardio room overlooking the pools.

With each passing year, more and more expansion plans are considered. In 2011, a 0.4-mile trail was built to offer views of downtown. A $4,000 grant from Wells Fargo and the hard work of the Rocky Mountain Youth Corps got the trail built. A deck on a previously unused hillside is a place where parents can relax and keep an eye on their kids in the pools. Yoga is also an option for the deck. Shaded gazebos were installed that can be rented out for private parties or those wishing to set up camp for the day. The hillside that has been developed was once considered a sacred area by the Utes, and the SSHRA was aware of this fact. During the trail building, the SSHRA worked to respect those areas. The SSHRA owns ten additional acres behind the current facility, and the future may see the walking path expanded even farther.

Bridges are just one of the enhancements to the Old Town Hot Springs that add to the beauty of the facility. © *Dagny McKinley 2012.*

Having a community space with an Olympic-size pool allows for athletes in the community to train and people to set and accomplish personal goals. Take, for example, John Ward, who swam one million yards in the Old Town Hot Springs pool. The *Steamboat Pilot* reported the feat:

> *Resident swims 1 million yards at Old Town Hot Springs*
> *By John F. Russell Thursday, March 18, 2010*
>
> *In May, longtime Steamboat Springs resident John Ward jumped into the lap pool at Old Town Hot Springs and began working toward a personal goal, 5,000 yards at a time.*
>
> *"I was just swimming laps, and I started doing the math," he said. "I figured that if I swam 5,000 yards for 200 days then I would reach a million before the end of the year. It's really not that much, but that's the goal I set for myself."*
>
> *Ward started working toward that goal in May and had reached his goal of 1 million yards by Feb. 14—well ahead of his May 2010 deadline.*
>
> *Diane Hagihara, the front desk manager at Old Town Hot Springs, said she sees a lot of people who come to the pool to swim laps on a regular*

basis but thinks Ward might be the first who has set his goal to reach the 1-million-yard mark—at least, he's the first she knows about.

She said Ward's dedication and consistency have helped him stand out at the health facility in downtown Steamboat Springs. His story came to light in December, after an employee asked him how often he swam, and he revealed his goal of making it to 1 million yards within a year. He was named member of the month at the Old Town Hot Springs that month, at which point he was 75 percent of the way to his goal.

"It was kind of a personal goal I set for myself," Ward said. "I only told one person, but it has kind of gotten out."

Ward's family moved here in the 1960s, and he grew up in Steamboat. He said he attended local schools and was a member of the swim team when he was younger. Ward will turn 50 in June, and he said he returned to swimming to stay in shape.

He said when he first started swimming, his goal was to make 10 laps across the pool. He gradually added more laps as he got into better shape, and he was swimming regularly when he set the goal of making it to a million yards in May.

Ward says he loves the Old Town Hot Springs pool and said the facility and friendly staff make getting to the pool and swimming a pleasure.

It takes him about an hour and a half to swim 100 laps, or 200 lengths, across the pool each day, which is 5,000 yards. He describes his style as slow but steady. He took a few cold days off this winter, but as he got closer to reaching his goal, he was motivated to get into the pool nearly every day.

"I swam more than a few days when the temperature was below zero," Ward said. "I wouldn't say I ran into any really big challenges, but there were a few times when it was hard to go down and get in the water."

Ward said he dropped a few pounds while chasing his goal but wasn't sure exactly how many.

"There were all kinds of positive things that came out of this," he said, "both mentally and physically."

In his spare time, he works as a J6 coach at the Steamboat Springs Winter Sports Club, where he teaches 8- and 9-year-old skiers the benefits of setting goals and staying in shape. He said his efforts have helped him keep pace with the young athletes who look to him for direction.

As far as future goals, Ward has at least one more.

"I want to go skiing in Chile," he said. "I'm not sure when it's going to happen, but that's my goal now."

Giving people challenges has been part of the community feel of the Old Town Hot Springs. For years, the Cardboard Classic on the ski mountain saw locals racing down Mount Werner in homemade cardboard vehicles that sometimes fell apart before crossing the finish line. Spectators and participants would throw snowballs and have a great time. But the SSHRA decided racing in cardboard vehicles shouldn't be limited to the ski mountain, so it created the Old Town Hot Springs Cardboard Classic. The challenge is to build a boat with cardboard, duct tape and Elmer's soluble glue that would float across the Olympic-size pool. While the challenge may sound like fun, there are some educational aspects to the creations. The goal is to have kids learn to work in teams, as well as understand the science behind floating and buoyancy. Cardboard paddles have been used, as have hands and feet for propelling. The length of the pool is twenty-five yards, and boat size is about eight feet long and three feet across. Two people ride in the boat. Awards are presented for fastest and most creative boats. The students really get into the spirit, as reported in the *Pilot*:

> *If for some reason the design didn't work for Charles Leech, Addison Sandvik, Eric Casey and Madeline Brusky, surely the name would give them an advantage.*
>
> *The four Soda Creek Elementary School fifth-graders were confident their cardboard and duct tape craft would cruise smoothly across the 25-yard lap pool during Old Town Hot Springs' third annual spring Cardboard Classic.*
>
> *If it didn't, the boat called* Cinatit *had some historical advantages.*
>
> *"We named it after the* Titanic,*" Casey said. "Except it's* Titanic *backward."*
>
> *Say what?*
>
> *"The* Titanic *sank and was slow," Sandvik said. "So we figured if we spelled it backward, ours wouldn't sink and it would be fast. It worked."*
>
> *The group outlasted competing teams from its own grade and was the fastest and most durable boat in the finals Friday…*
>
> *So what happens next with the* Cinatit*? The original plan was to burn it. But unlike the original* Titanic, *the* Cinatit *will see more than just its maiden voyage.*
>
> *"We might put it in the* [winter] *Cardboard Classic," Brusky said.*

Another way the springs connect with the younger generation is through the Cool Pool Pentathlon. Teams of five climb stairs, rush down the slides, run through

Female bathers enjoy the Old Town Hot Springs. *Courtesy Steamboat Springs Health and Recreation Association.*

the shallow pool, kayak the lap pool and swim or kick one length of the lap pool. Near Halloween, the Pumpkin Float gets kids and pumpkins into the pools.

The SSHRA also hosts an annual off-site event called the Open Water Swim at Bald Eagle Lake. Kids eleven and older race in open waters, helping kids get ready for triathlons, train or just get fit.

Despite the world-class facilities of Steamboat, this town is in many ways still a small one. Families help families in need. Businesses and organizations are always willing to help out. Such was the case when Lindsay Bathke discovered a swollen blood vessel the size of a golf ball in her brain. In true Steamboat tradition, her brother and friend set up a fundraiser called Laps for Lindsay. The goal was to get pledges for swimming laps. Donations were welcome and food, prizes and other fun activities were offered. The event was a success. Events like these occur more frequently than not in this town, with everyone pulling together to help families fund overwhelming medical expenses, grieve in tragedy and celebrate good times.

As far as good times are concerned, even the dogs of Steamboat are included. Poochy Paddle is an annual event that lets Steamboat's dogs enjoy the hot spring. Dogs are invited for a swim or a soak, to jump, shake and splash as much as they like before the pools are cleaned out in the spring. Owners pay a nominal fee, and then the fun begins. Dog owners bring extra food for dogs and cats that is donated to the local animal shelter and animals in need. Over one hundred dogs have been found carousing and splashing at one time. The event usually takes place on the closing day of the Steamboat Ski Area. It's a great event for people who want to get away from the mountain, and it also helps prevent thousands of dogs from showing up on a single day, as many owners are skiing that last day. As locals often say, there's two dogs for every person in Steamboat.

With more dogs and more people, another expansion is in the works. Due to public desire and an eye on the future of tourism, the SSHRA is considering expanding the building and fitness facilities, including a bigger spa and fulfilling the town's need for an indoor climbing wall.

People have always been drawn to the healing powers of water for physical therapy, fitness or pure enjoyment. The Old Town Hot Springs provide a place where all these needs can be fulfilled.

Chapter 9

Soda Spring

Mineralogy of Spring, Early 1900s

Temperature 58 degrees F

	Parts in 100,000
Chloride Sodium	38.01
Chloride Magnesia	20.42
Chloride Potassium	9.16
Bicarbonate Soda	202.45
Bicarbonate Lime	73.84
Bicarbonate Iron	0.50
Silicia	2.88
Total Solids	347.26

Mineralogy of Spring, 2012

Temperature	55 degrees
Alkalinity, mg/l	1020
Bicarbonate, mg/l	1020
Boron	none

Calcium, dissolved, mg/l	24
Carbonate, mg/l	none
Chloride, mg/l	378
Coliforms, -/100 ml (1/100 ml)	none
Fluoride, mg/l (4 mg/l)	1.0
Hardness, mg/l	269
Magnesium, dissolved, mg/l	51
Nitrate, dissolved, mg/l	.46
Nitrate/Nitrate, mg/l	.46
Nitrite, dissolved, mg/l	none
Nitrogen, mg/l	none
Ph, units	7.7
Potassium, dissolved, mg/l (5,600 mg/l)	.53
Sodium absorption ration	18.83
Sodium, dissolved, mg/l (3,000 mg/l)	702
Silica, mg/l	none
Sulfate, mg/l (250 mg/l)	196
Arsenic, dissolved, mg/l (0.05 mg/l)	none
Barium, dissolved, mg/l (1.0 mg/l)	.04
Cadmium, dissolved, mg/ (0.010 mg/l)	none
Iron, dissolved, mg/1 (0.3 mg/l)	1.96
Lithium, dissolved, mg/l	.85

mg/l = milligrams per liter
Elements: Maximum recommended by state and federal government in ()
*Source: Tread of Pioneers Museum, "The Springs of Steamboat Springs Walking Tour"

LOCATION

The Soda Spring can be identified by the pavilion housing the spring in the West Lincoln Park, directly across from the Bud Werner Memorial Library. The spring sits on the southwest corner of 13th and Lincoln Avenue (Highway 40).

What to Expect

While the pavilion has been kept in good condition and is the original structure built to house the spring, the Soda or Pavilion Spring is now less than a puddle in the center of the pavilion often filled with leaves and other debris. After reading the history of the springs, the Soda Spring was the one spring I was really excited to experience. I imagined sipping the naturally carbonated waters that had been such a pleasure to so many and maybe making my own soda. I wondered why the waters hadn't been bottled and sold. I wanted the clean, refreshing taste. I practically ran to the spring.

When I looked in, there was nothing but a small dab of water in the middle of a circle. I looked around and thought perhaps the flow had been temporarily stopped because of construction going on across the road. When I realized the flow had been stopped for good, my heart was heavy. I wondered how something that had refreshed so many had been allowed to slip away underground. I feel that somehow we, as people, failed the spring and failed ourselves by letting it go. This spring used to be the first stop for many travelers coming to town and might still be flowing as it once did if not for human error.

History

The story of this spring is a sad one, for not much remains of this legendary spring. However, the tale must be told and can perhaps be a model for future decisions regarding natural treasures in this town and others.

The Soda Spring is also known as the Pavilion Spring for the pavilion constructed over it by the early settlers. This spring was the most popular drinking water of the group of springs in Steamboat. Long before there was soda, the natural effervescence of this spring lured settlers and travelers alike. Until the 1980s, travelers and old-timers would stop for a glass of fizzy lemonade on their way through town. Other flavors of pop were tried, but the lemonade was always the favorite.

> *Recipe for Steamboat Fizz:*
> *Add two tablespoons of lemon syrup to a cup of soda springs water, stir and hand to young lady waiting.*

In 1905, a traveler encountered and was won over by the Soda Spring, as seen in this piece from the collection of the Steamboat Spring High School Library:

> *PAVILION SPRING*
> *The analysis here indicates a chalybeate water, if judged by taste and effect, has no superior in the world. Its mineral constituents are so consistently mixed and modified, that it can be, and is constantly and daily used. Dr. Denison, a well-known and eminent physician of Denver, and also a learned writer on the "Mineral Waters of Colorado," calls attention to this delicious and valuable water, and to the great quantity that can be assimilated by the drinker. It is sparkling and highly effervescent, from the large quantity of carbonic acid gas contained. It requires a strong effort of the will to abandon to use of this water, and it is destined to be used in all parts of the civilized world. It is an active remedy where there is surplus of acid, or catarrhal condition of the digestive canal. There is a sensation of vigor, or exhilaration, follows the drinking of this water. It is a natural champagne, perhaps superior in richness and delicacy of its taste to Giesshubler or Apollinaris.*

Another cheekish account recalls, "Humans…in the manner of the times erected a handsome pavilion to enhance their enjoyment of its healthful effects. It is not recorded whether the waters of this spring found additional practical uses over on the Brooklyn side of the Yampa where more bracing refreshments could be obtained." Brooklyn, across the bridge from downtown Steamboat Springs, was the bawdy neighborhood. When Steamboat was a dry town, Brooklyn offered prostitution and alcohol. The women of those institutions used to parade down the Main Street of Steamboat to solicit business.

The Crawford family, the first white settlers of Steamboat, enjoyed the clear, sparkling waters of this spring as much as anyone. Even two generations later, granddaughter of the founding family, Lulita Crawford, fondly remembered the refreshing waters. In a letter to Max Selbe in 1967, Lulita recalls:

> *We used to enjoy the Navajo* [Spring] *water because it was cold and sparkling. Because it had some sulphur taste, we never liked it quite so well as the Soda Spring, which was our all-time favorite. The Soda Spring is still our favorite. I expect Margaret and I are about the only people who*

The pavilion is all that remains of the once famous Soda Spring. © *Dagny McKinley 2012*.

ever drink from it these days. The first stop we make when we drive into Steamboat is at the Soda Spring to dip up some bubbles. And we love lemonade made with the soda water.

In the late 1800s and early 1900s, tourists flocked to Steamboat for the healing powers of the waters. The Cabin Hotel overlooked the Pavilion or Soda Spring and was host to those looking to cure their ailments. Touted as the "finest hotel in northwest Colorado," this grand structure offered weary travelers a choice of one hundred rooms, access to waters to soak in and mineral waters to sample for their health. Guests had only to walk a few feet out the front door of this building before they could drink soda water fresh from the ground.

Everyone's hope in Steamboat was that with the arrival of the train, the town would thrive. The transportation brought exposure until the Great Depression swept the country, effectively killing the business of the hotel. After struggling for a couple years, the hotel burned down. Perhaps the Great Depression is part of the reason the Soda Spring never attained the worldwide fame so many people thought it would. The Soda Spring was of the same ilk as Perrier.

Despite modern-day warnings that the Soda Spring could contain bacteriological contamination, old-timers regularly drank from the spring anytime they came into town. According to an article in the *Steamboat Pilot*, Margaret Herbert, a third-generation native, made lemonade from the spring water all her life. In the early part of the century, prior to soda pop, the fizzy water was extremely popular. "The spring has lost some of its bubble, says Herbert, but it still has a sparkly lemon taste that makes you belch," reported the *Pilot*.

Gloria Gossard, in an interview with the *Pilot*, remembers people taking Soda Spring water over to the Cabin Hotel to add a squeezed lemon. Even after the hotel burned down, the springs remained part of the local culture.

When Don Comb was a kid in the 1930s, he was paid twenty-five cents a week to send a gallon of the Soda Spring water to a dentist in Denver because the dentist liked the taste of the water. Comb also worked on the original Environmental Impact Statement for the highway widening project in 1974. By 1988, the City of Steamboat Springs was ready to expand the highway. While the flow of the Soda Spring had lessened by 1988, it was nothing compared to what was about to come.

With a population increase, traffic began to back up, and the town felt Lincoln Avenue needed to be widened to accommodate the increasing flow

Ladies at the Soda Spring. The Cabin Hotel in the background. *Courtesy Tread of Pioneers Museum.*

through town. In order to expand the highway to four lanes, the Soda Spring would have to be moved.

Locals like George Tolles, whose name is linked to the mineral springs of Steamboat for his championing of their benefits, knew that moving the waters would destroy them. Other residents shared his concern and referenced the famous Steamboat Spring that was silenced when the railroad blasted through the hills nearby to lay tracks. Officials assured the public that there would be no damage done by moving the spring and that they would take all necessary precautions to ensure this historic spring's flow wasn't interrupted.

Construction began in July 1988. The Soda Spring was capped with a spring box and piped through plastic PVC piping in an effort to ensure the water "remain drinkable and should also retain its fizz," according to the *Steamboat Pilot*'s "It Won't Hurt to Move Spring, Officials Say." A manhole covers the spot where the spring used to flow and is supposed to allow access to the original bubbling vent. The state paid the cost of moving the gazebo and piping the spring into it. The cost to expand the highway and move the spring ran approximately $1.5 to $2 million and was funded by the State of Colorado. The expansion paid for jobs.

An interview in the *Steamboat Springs Review* quoted Steamboat Springs public works director Kirk Madsen, who boasted that moving the spring might not be "as difficult as imagined."

However, previous events should have foreshadowed what would happen to the Soda Spring. The two-lane highway was originally paved in 1946. At the time, workers found the ground pocked with springs and sinkholes. Others working on the expansion admitted to drainage issues. Lulita Crawford Pritchett voiced the sentiments of many in Steamboat when she addressed the chamber of commerce, the city council and the people of Steamboat Springs in an open letter that reads:

> *I have always been proud of Steamboat Springs. It goes against the grain to apologize for the town, but this morning I did. As I was walking to the Soda Spring with my big cup, a tourist stopped his car to ask, "where can we find a spring we can drink from? We have looked at several, but—"*
>
> *I remember how the old Town Company used to work with the springs to maintain them. How well, I know the labor and money that went into keeping them clean and attractive!*
>
> *The mineral springs were the reason my grandfather founded this town. They should be kept up if for no other reason than they are here. They are the town's signature.*
>
> *Though all the springs have curative properties, the Soda Spring (which old timers called the Iron Spring) was always best for drinking. People drank it for pure pleasure. Ice cold, carbonated, it made lemonade which would put modern bottled pop to shame. Generations of Steamboat people grew up on it.*
>
> *Originally the Soda (Iron) spring had a strong flow. It bubbled boisterously. When a storm was coming, it boiled out of its rock basin. The flow was cut when the present road was blasted. Though the spring still bubbles lethargically, most of the water apparently drains into the river.*
>
> *I am told this spring is again threatened by the projected highway. Unlovely as the spring is in its present state of neglect, I strongly urge that it not be forfeited. If you modern dwellers in Steamboat had enjoyed it as did the old timers, you would know it is worth fighting for. Why not try to find the original flow?*
>
> *Vandals have ruined the flagstone around the Sulphur Spring, and the mineral waters themselves have eaten away at the concrete. But funds are*

After the city moved the Soda Spring to expand the highway, the once-famous carbonated spring was reduced to a trickle. © *Dagny McKinley 2012.*

> *always found for town maintenance, road work, ski projects. Why not for maintenance of the springs?*
>
> *Why did the town managers allow toilet facilities to be built so close to the Sulphur Spring?*
>
> *The Lithia Spring is still there across the railroad tracks, but the sign at the road turn-off has been knocked down. As I walked past the remains of lumber operations, I heard one killdeer where many used to nest among several smaller springs. A house overshadows the big spring. Only one other spring—in Germany—compares with this, we used to be told.*
>
> *In the loud demands of the present, it is easy to sacrifice priceless natural treasures and beauty that can never be regained.*
>
> *We have already wasted too much of our heritage. Let's not waste all of it.*

As the piping of the Soda Spring began, the spring was lost. Perhaps the spring decided that if it had to be moved, it would rather not flow. Today, not even a few bubbles can be found.

Across the park and down the bank toward the Yampa River, there is runoff from a carbonated spring that may very well be the Soda Spring.

The water is cool and bubbly. Take a glass, enjoy a sip or two and imagine what it might have been like to see this water bursting from the ground during a storm.

Chapter 10

The Steamboat Spring

Mineralogy of Spring, Early 1900s

Temperature 80 degrees F

	Grains in One Pint
Chloride of Sodium	14.62
Sulphate Potassa	0.61
Sulphate Magnesia	0.85
Sulphate Lime	1.57
Carbonate Soda	5.48
Phosphoric Acid	Trace
Silicic Acid	Trace
Iron Oxide	Trace
Total Solids	23.13

	Cubic inches of gas
Hydrosulphuric Acid	6.14
Carbonic Acid	0.90

MINERALOGY OF SPRING, 2012

Temperature	79 degrees
Alkalinity, mg/l	2425
Bicarbonate, mg/l	2425
Boron	none
Calcium, dissolved, mg/l	24
Carbonate, mg/l	none
Chloride, mg/l	1470
Coliforms, -/100 ml (1/100 ml)	none
Fluoride, mg/l (4 mg/l)	3.5
Hardness, mg/l	187
Magnesium, dissolved, mg/l	31
Nitrate, dissolved, mg/l	none
Nitrate/Nitrate, mg/l	none
Nitrite, dissolved, mg/l	none
Nitrogen, mg/l	none
Ph, units	7.9
Potassium, dissolved, mg/l (5,600 mg/l)	130
Sodium absorption ration	79.45
Sodium, dissolved, mg/l (3,000 mg/l)	2470
Silica, mg/l	none
Sulfate, mg/l (250 mg/l)	628
Arsenic, dissolved, mg/l (0.05 mg/l)	.110
Barium, dissolved, mg/l (1.0 mg/l)	.03
Cadmium, dissolved, mg/ (0.010 mg/l)	none
Iron, dissolved, mg/1 (0.3 mg/l)	none
Lithium, dissolved, mg/l	3.75

mg/l = milligrams per liter

Elements: Maximum recommended by state and federal government in ()

*Source: Tread of Pioneers Museum, "The Springs of Steamboat Springs Walking Tour"

Location

The Steamboat Spring sits just above the Yampa River and below Black Sulphur Spring. Parking can be found at the Bud Werner Memorial Library or at the Art Depot across the street. To get there from downtown Steamboat, turn left on 13th Street. Cross the bridge and park at the Art Depot on the right. Walk across the road and you will find the Black Sulphur Spring and the Steamboat Spring.

What to Expect

The Steamboat Spring was believed to be Colorado's only geyser. This geyser once shot water up to fifteen feet in the air. Today, there are two small bubbling sources. I was prepared for the Steamboat Spring. I had heard the stories about the spring being silenced. As I looked into the depths of the spring, I wondered how grand it might have been to see that water shooting up, to listen to the internal churning sounding like a southern steamboat chugging down the river. I was not able to find any images of the spring shooting into the air. There are written records and recorded memories. Today, there is a viewing platform where visitors can visit the spring and imagine what once was.

History

The name "Steamboat Springs" has been associated with this area since 1839 or possibly even before that. The earliest known written reference to the name is found in the journals of Frederick Wislizenus, a German doctor who visited the spring in 1839 after accompanying a fur-trading caravan. By that date, the Rocky Mountain fur trade had been thriving for seventeen years. Wislizenus's journal was published in 1912 by the Missouri Historical Society and titled *A Journey to the Rocky Mountains in the Year 1839*.

Wislizenus wrote:

> *A warm spring…deserves special mention. The spring issues from a block of lime, which it formed itself in the course of time. The stream, as thick as an*

> *arm, spouts out in abrupt pulsations, and runs into the river of the rock, which is coated with the oxide of iron and white crystals of salt. About six feet off are two smaller openings, one of which is obstructed, while the other is still open.*
>
> *From the latter there issues with puffing noise...a gas mingled with vapor. This has a somewhat pungent and benumbing odor...The puffing noise deceptively resembles the well known sound of an engine, for which reason* [it] *is also known by the name "Steamboat." Gladly would I have spent some time in this most interesting valley, but my companions insisted on pushing on.*

Steamboats and the sounds they made were well known in the West by the 1830s; the first steamboat floated the upper Missouri River in 1819.

Another reference to the Steamboat Spring is found in the journals of George Augustus Frederick Ruxton, a young British adventurer. His *Adventures in Mexico and the Rocky Mountains* was published in 1847 in London. His experience with the spring was primal:

> *The American and Canadian trappers assert that the numerous springs which, under the head of Beer, Soda, Steamboat...abound in the Rocky Mountains, are the spots where his satanic majesty comes up from his kitchen to breathe the sweet, fresh air, which must doubtless be refreshing to his worship after a few hours spent in superintending the culinary process going on below.*

When James Crawford decided to stake his claim to the land, he pinned his claim on a quaking aspen just a few feet away from the Steamboat Spring in 1874. The tree and the spring remained until the railroad was graded; the tree was cut down in 1908.

Crawford recounts the story of the first men to name the spring in Lulita Crawford's *Cabin at Medicine Springs*:

> *"Medicine Springs" was the Indian name for the mineral pools near which Pa had built his cabin. "Steamboat Springs" was the white man's name and this was rooted in the dim era of the early fur trappers. Pa had learned the tale from some old mountain men—how two French trappers, coming along Bear River for the first time, had heard a chugging sound. Believing they had reached a navigated river, one had exclaimed to the other, "Steamboat by gar!" They had soon discovered that their find was no steamboat—only a spring, spouting noisily to the south of the river and hurling water like a small geyser.*

Later in the season, it could be heard almost to the Crawfords' cabin.

A commemorative steamboat was built in the West Lincoln Park in tribute to the sound of the spring the town was named after. © *Dagny McKinley 2012.*

The fumes from the spring were never much considered by the Crawfords until the morning James Crawford woke up to find his new red wagon box gone. He had left the wagon box on the hillside near the chugging Steamboat Spring:

> *The next morning he looked across the river and the red wagon box was gone and a grey one was in place. James said "Maggie" he snorted "somebody's stolen our wagon." He leaped into the river and crossed to the far bank. A few minutes later he came back with a sheepish grin on his face. "It's our wagon all right," he said, "but it has changed color. I must have left it too close to the springs. The only thing I can figure is the mineral in the air made red paint turn lead color.*

A later analysis from the collection of the Steamboat Springs High School Library proclaimed:

> *This spring is situated just out of view, to the right, in the accompanying picture of the foot-bridge. It bursts out, in irregular gushes, from a ragged*

> *orifice in the solid "rock in place." The quantity of carbonic acid gas is so great and is sent up from its subterranean place of production under such pressure as to eject the water through the outer aperture with almost an explosive violence. The interesting phenomenon here witnessed, is the water coming forth, not in a continuous flow, but in pulsations accompanied by a pulling sound, resembling that made by the exhaust pipes of a steamboat as it approaches the wharf. This peculiar feature, which gives to this spring and the town their name, is probably due in part to the pressure and partly to the torturous windings and vacuous cavities that the water traverses in its passage to the surface. The analysis shows an alkaline-saline water so rich in carbonic acid gas that no counterpart is to be found in any land. This water will be bottled without the aid of machinery otherwise than an iron pipe, the natural pressure doing all the work.*

For fun, kids would throw rocks into the Steamboat Spring to see them tossed back out. The rocks that were thrown in partially clogged the flow of the geyser, and before the railroad made its way into town, the spouting had diminished slightly.

This famous spring was part of the pride of the town. Yet having the railroad make a stop here became a top priority. The citizens of Steamboat raised $50,000 when the railroad company threatened to bypass Steamboat in favor of an easier route.

There were many challenges of laying a railroad in this rocky terrain. As the railroad company neared the Steamboat Spring, it began blasting rock through the hill that rises behind the spring. Time and money were the concerns, not the natural treasures of the towns it was working through. The *Steamboat Springs Review* recounts what it imagines that day was like in its August 15, 1984 issue:

> *THE DAY THEY TAMED THE STEAMBOAT SPRING*
> *The morning of June 25, 1908 dawned clear and brilliant in Steamboat Springs. The town of less than 1,000 people. There was a certain excitement in Steamboat Springs that summer. An excitement that was carried on the winds of change, progress and promise. For the railroad was coming through town for the first time, and the work crew which slowly snaked its way along the southern bank of the Yampa, clearing the right of way and laying track was a constant reminder of big things to come. Before long, it was thought, Steamboat Springs would be hosting thousands of vacationers arriving by rail. Politicians and businessmen*

The railroad cutting through to Steamboat, circa 1908. *Courtesy Tread of Pioneers Museum.*

were boasting of the boom that would come with the long-awaited link-up with Denver.

That particular morning the railroad construction crew was faced with the formidable task of blasting the roadbed through a sheer sandstone formation which stood just above the famous Steamboat Spring at the west edge of town. The crew, preparing to dynamite the hillside, worked steadily burying charges and laying fuse wire.

At noon, the workers broke for lunch, sprawling about on the rocks with sandwiches and cookies. As they ate, the men took in the mid-day performance of the incredible Steamboat Spring as it gurgled and chugged, spouting a geyser of hot mineral water first 5 feet, then 10 feet, then an astounding 14 feet into the air, then falling momentarily silent as it rebuilt pressure for its next surge. Crew members, some who had been working on the line all the way from Denver, marveled at the natural phenomenon which had given the town its name. Most had never seen anything like it in their lives.

After lunch, the workers cleared the area as the foreman readied to light the first fuse. The flame smoked and hissed its way across the rocks, finally reaching the charge, sending a shower of rocks and dirt into the river with a resounding blast. A second fuse was lit, then a third, with accompanying explosions and showers of debris. When the dust cleared, the crew moved in to begin removing loosened boulders and piles of broken sandstone. But as they worked it became apparent something was very different.

The Steamboat Spring, an active thermal geyser since perhaps the dawn of man, had stopped chugging. Upon closer examination, the workers noted that a large crack had formed in the rock that encircled the spring, allowing water to gush out in large volumes, splashing down the bank into the river in a steady stream. Two crewmen yelled to the foreman that the spring had been damaged. He waved them back to their shovels. Work on the railroad continued.

A few townspeople expressed their dismay that the namesake landmark had been destroyed. But the festive mood created by the new railroad and its promise of tourist wealth soon silenced all misgivings over the loss of the Steamboat Spring.

It is grossly ironic that today the railroad carries no tourists and the town of Steamboat Springs is still trying to figure out how to attract vacationers, with proposals for new malls, golf courses, race tracks and the like. Our modern developers should all be required to spend a silent moment at the site of the old Steamboat Spring—and then instructed to proceed with great caution.

A bubble, all that's left of the once famous geyser. © *Dagny McKinley 2012*.

The disappearance of the spring was something that was remembered by all who had known the spring, and blame was passed around, as seen in this letter from George L. Killion, chairman of the board of MGM Studios in Los Angeles, to his childhood friend Celly Merrill:

Dated November 2, 1970

To. Mr. Celly Merrill

Dear Celly:
It has been a long trail since those days in Steamboat Springs, but despite the intervening years I recall those carefree days you, Maury Leckenby and I spent in and about our beloved Steamboat Springs. I don't recall that I am responsible for throwing the big rocks in the mineral springs. I always thought it was you.

Before his passing, Maurice Leckenby visited me two or three times here in California. My father, James Killion and Maury's father, as you know, were old pioneers in Steamboat...

I would like to visit Steamboat again, and I shall some fine day.

Sincerely,
George L. Killion.

In 1959, town officials tried to restore the chugging sound of the famous spring by building an acoustical rock formation around it. The chemicals around the spring that once ate the paint from James Crawford's wagon ate away any possibility of restoration. Cleaning the spring did nothing to bring it back. To this day, the spring remains silent.

Chapter 11

Strawberry Park Hot Springs

Mineralogy of Springs, Early 1900s

Hot spring
Temperature 152 degrees F
Flow 2700 gallons per hour

	Parts in 1,000,000
Sodium Chloride	442.40
Calcium Chloride	171.93
Sililic Oxide	92.40
Calcium Sulphate	89.00
Calcium Carbonate	84.22
Potassium Chloride	31.89
Barium Carbonate	26.80
Magnesium Carbonate	27.58
Total Solids	996.22

Hot Springs Mineralogy, 1988

	mg/l
Arsenic	0
Boron	.28
Cadmium	0
Calcium	7.8
Chloride	130
Flouride	17
Iron	.11
Lithium	0
Magnesium	0.1
Manganese	.01
Mercury	0
Nitrogen	.02
Phosphate	.05
Potassium	8.8
Selenium	0
Silica	89
Sodium	160
Sulfate	45
Zinc	0

Alkalinity
As Calcium

Carbonate	114
As Bicarbonate	139
Hardness total	20
Specific conductance Micromohs	1,000

Total dissolved solids	527
Ph	7.8
Discharge	35 (Gpm)
Temperature	147

Location

Strawberry Park Hot Springs is located seven miles outside the township of Steamboat Springs. To get there coming from Mount Werner, take Lincoln Avenue and turn right at the post office or 3rd Street. At the first stop sign, take another right onto Fish Creek Falls Road. A short way up, you'll see a sign directing you to take a left to the hot springs (Amethyst Road). You'll follow this until the road becomes Strawberry Park Road or County Road 36 and is unpaved. From here, you'll need either four-wheel drive or chains in the winter and spring. Follow the dirt road up a hill and over a few bumps until the road dead-ends into the parking lot of the springs.

What to Expect

The road to the springs was once traveled by horse, then skis and now by car. Longtime locals swear the road was in better condition during the days of skiing than it is today.

What you experience will depend on whether you go up during the day or at night. A daytime visit means you can visually enjoy the natural landscape. Aspen trees and pines surround the springs that are set into a jeweled valley. The owner, Don Johnson, has developed the area into three main pools and two private pools. Each of the pools is temperature controlled, one a little cooler than the next. A few years back, the river flowed adjacent to the main pools, and one could jump in the frigid waters in winter or summer to cool off. Today, the river is mixed with warm water for a less shocking experience.

There are also private pools, some accessible for guests and some for those with appointments for water therapy. Cabins dot the land past the springs and at the end of the hot springs hiking trail, where the lodgings are rustic but beautiful. Covered wagons and camping spots let summer visitors enjoy some solitude. Lodging comes with a barbecue to cook and fuel for the barbecue, but guests must bring all their own utensils, sleeping gear, food, etc. The hot springs are off-grid. Solar energy supplies the power.

An alternative to driving to the springs is to hike in. To get to the hiking trailhead, follow Lincoln Avenue north out of town. Turn right on Elk River Road (at the 7-Eleven). Follow this road for a few miles and beware of the one sharp turn to the right. Travel a little farther until you see the signs for the Mad Creek Trailhead. (This parking area is not plowed in the winter, so if there is heavy snow, there won't be a place to park.) Park at the trailhead for

The earliest mode of transportation to the hot springs was by horse, then carriage, skis and finally car. *Courtesy Tread of Pioneers Museum.*

Mad Creek. Hike back down the road the way you came for a few hundred feet, cross a bridge and keep going until you see a pathway leading off to your left. You must go through a gate to stay on the trail, but remember to lock the gate behind you as cattle graze this area in the summer. A 3.5-mile hike in will take you slightly uphill alongside a creek. The views down valley are beautiful. Don't miss the tree with the animal skulls and art installations hanging from it or the old pioneer cabins along the way. Dogs are welcome on the trail but not on the hot springs property.

At night, everything is different. Stars shine overhead, or a slight snow may fall. Shapes illuminate and disappear. Those under eighteen are not allowed in after dark for safety reasons. Flashlights are recommended to navigate down to the springs. Only shine lights at the ground and not at the surrounding pools to respect people's privacy. Skinny-dipping is allowed after dark, but once in the water, it's impossible to know who is dressed and who isn't. Feel around the pools, as your feet sink into gravelly softness. Little pockets of heat, where water seeps up, can be found. The heat causes a pleasant burning sensation on the toes.

Bring water or Gatorade for hydration. No alcohol is allowed at the springs. Shoes with good tread are recommended for winter when the walkways are often icy or slushy.

History

The name Strawberry Park Hot Springs comes from a bit of confusion. The signs heading out toward the springs read "Buffalo Pass, Strawberry Park, Hot Springs." Although the hot springs are not in Strawberry Park, too many people lumped the signage for Strawberry Park and the hot springs together until the name finally stuck.

Scientific analysis has shown the water from this spring to be almost chemically pure. The beauty of the surroundings has a strong draw for people. Inevitably, if asked for a story about the mineral springs in town, locals will have a story about the Strawberry Park Hot Springs.

Like many of the springs in Steamboat, the Strawberry Park Hot Springs have undergone a few name changes during their history. From the hot springs to Big Hot Spring, Boiling Spring, Mighty Fine Spring and Routt Hot Spring, this body of water has also undergone several physical transformations. With the hottest water in town, the springs were a draw for early settlers on an outing, as seen in the following from the collection of the Steamboat Springs High School Library:

> *Big Hot Spring*
> *This remarkable creation is situated several miles from Steamboat, and at an increased elevation of about 300 feet, discharging through a tissue in the granite rock not less than 400 gallons per minute, having a temperature of 150 degrees F. Eggs are soft boiled in the most perfect manner, tea steeped, game and fish cooked and various other culinary performances are accomplished by this naturally heated water. This water is to be carried down in pipes, under a great pressure and used in connection with those of the main group, making an incomparable quantity of high temperature*

An early day at the Strawberry Park Hot Springs. *Courtesy Tread of Pioneers Museum*

> *mineral water. Dead snakes are often seen along the edges of the flow, having been actually cooked by accidentally getting into the water, or attempting to cross the hot brook. The line of contact between the granite rocks and the sedimentary is near by, also an outcrop of eruptive rocks in the vicinity, which reveals, to some extent, the origin or cause, of such highly heated water. This water is soft, containing little or no lime, evidently pursues its course granite only from the natural laboratory that heats it.*

Sources are conflicted as to where or how the heat is generated for these springs. Popular belief once held that Crystal Peak, now named Rocky Peak, an acidic volcanic rock, heated the springs. Others say Hahn's Peak is responsible. Five hot springs compose the Strawberry Park Hot Springs.

Just as the Crawford family were intimate with the springs close to home, they were also intimate with the Strawberry Park Hot Springs, and their account should be noted:

Five women boiling eggs in the hot springs, August 3, 1907. Boiling an egg took ten minutes, and the eggs never hard-boiled, *Courtesy Tread of Pioneers Museum.*

The hot springs [are located] *6 or 7 miles north of town. These springs lay in the roughest kind of country. At the bottom of a deep crease between steep hills ran the creek and from either side dozens of rivulets drained steaming water into it. The water was almost boiling. In summer, Pa liked to show off the Hot Springs to visiting friends and relatives and Ma always took eggs along to cook in the hot water. With a dipper she would nest them in the sand at the bottom of a bubbling pool, and in half an hour she would lift them out nicely coddled. No matter how long they stayed in the water they never got hard boiled.*

Here the creek never froze. For ¼ acre there was no snow. In cracks the rocks grass grew as green as in May and yellow monkey flowers bloomed.

Men cooking sausages in the hot springs, perhaps after a day of hunting. *Courtesy Ruth B. Perry and Strawberry Park Hot Springs.*

Logan Crawford, in *Remember the Old Yampa Valley*, recalls one adventure when he led some explorers into the hills while he pursued an old vendetta against a huge bear named Old Mose. After a rough chase and despite not catching the bear, Logan was still inspired by the beauty of his surroundings: "Steam from the spring had draped long festoons of crystals like glorified popcorn strings upon Christmas trees. Only this little spot in Routt County had diamonds, rubies, emeralds and sapphires hanging from trees. And yellow monkey flowers blooming among rocks in the winter!"

The Recreation Edition of the *Steamboat Pilot* reported in 1940:

> *The beauty of the scenery, the grand mountain peaks and the clear stream of pure, cold water that flows past the springs, all contribute to the great possibilities of the proper development of these springs. A favorite pleasure trip to townspeople and visitors is to put up a lunch basket, place a few spoonfuls of tea in the teapot, take a few fresh eggs and motor to the Hot Springs, drop the eggs into the hot spring, pour the hot water in the teapot and in ten minutes have a hot meal cooked by radium heated water…The remarkable pain-relieving effects of natural hot baths bear a close resemblance to the similar power of the X-rays, which contain most of the radium emanations. The water of Bath, Vichy, Flombleres and Baden-Baden have been found possessed of marked radio-activity by Sir William Ramsey, Professor Dewar and others, while a similar quality has been found in the waters of Hot Springs, Ark., by Professor Boltwood. If this be the secret of their mysterious power, Steamboat Springs would rank high among them.*

Today, moose, elk, deer, foxes and, on rare occasions, mountain lions have been seen, along with ermine, marmots, pine marten, chickadees, steller's jays and American dippers, according to Joe Stepan, manager of the Strawberry Park Hot Springs.

When the Steamboat Springs Health and Recreation Association (SSHRA) owned the springs, along with the Old Town Hot Springs, they remained undeveloped, and the natural setting invited good and bad use of the area. These hot springs ran down into knee-deep pools surrounded by a ring of rocks. In order to cool off the pools, bathers used a wooden irrigation gate to let in cold water from the Hot Springs Creek. Dogs, friends and horses could all be found frolicking by and in the springs.

In the 1950s, Tom Wither and his family went for picnics every Sunday after church. Sometimes the picnics took them up to the Strawberry Park Hot Springs in the summer when the pools were still very primitive. They always brought eggs. There was a specific place, just down the hill as you walked toward the springs, where eggs were left to soft-boil. The water was very hot right there, and the area was only about two feet in diameter with rocks built around it. The Wither family would put the eggs in when they arrived, and in an hour or so, or by the time they were ready to leave, they would rush to see whose eggs had soft-boiled.

Tom's childhood created an appreciation for the springs. He spent time there in the hippie era and, as an adult, cross-country skied in. One time he

Strawberry Park Hot Springs before they were developed. People used to create private baths by moving the rocks around. *Courtesy Bill Fetcher.*

made it in with a friend for an afternoon soak. After the sun had set, the road turned to a sheet of ice; they slid and skidded until they had to take their skis off and walk out. Around 2002, Tom and his partner were up at the springs. Tom's partner had never been to the springs. They had a wonderful soak, and then the sun set around 4:00 p.m. The roads turned to a sheet of solid ice. But Tom had a Subaru. As his partner was driving up a hill, a person driving in the opposite direction warned them to be careful going up the hill because somebody down below had an accident. They weren't even a mile from the hot springs. Tom remembered:

> *Gary was driving and we just lost it, so I said go for the snowbank, so we did a 360 and went into the snowbank. Thank goodness we did; otherwise we would have crashed into this car. We got out of the car and we each helped the other people get out, but Gary fell on the ice and broke his wrist. He does hair for a living, so that was kind of a bad memory for him. But overall it was a wonderful experience.*

Rick and Linda Banter fondly remember their visits to the springs before the SSHRA sold the property to raise money to develop the Old Town Hot Springs:

People once skied to the hot springs. Now, people can drive with four-wheel-drive vehicles or vehicles with chains. *Courtesy Ruth B. Perry and Strawberry Park Hot Springs.*

> *Sometimes we would take a tent and camp downstream where the water was cold. Where the water comes out of the hillside, you would mix it with the creek. Everybody that was there would take rocks and make a pool and go as far upstream as they wanted. If you went upstream where the water was coming out, it was almost boiling. You couldn't even go in there. If you went downstream you could temper the temperature because the cold water mixed with the boiling water coming out of the hillside. So everybody would make their own little pools and just sit and luxuriate in their pool. There would be dogs. People would come up on horseback. It was quite a lovely scene. I would camp overnight. It was a very, very friendly scene, and people really enjoyed it.*
>
> *And then I think what happened was in the late '70s, early '80s, there was an article in the Denver paper about it and people started coming up. They were not treating the property properly—there was garbage and stuff—and that created the situation where the owner of the land closed it up and made it BLM* [Bureau of Land Management] *land.*
>
> *Every year there was the Cardboard Classic on the ski mountain. The year the springs closed, well, a friend of ours who owned Norm's TV at the time that then became Yampa Valley Sound decided to show how we locals felt. The city was going to make a deal to make the property a little more commercial—it was a sad moment for most of the locals. Greg Goosely and John Sendack built the Hot Springs Heartbreak Hotel out of eighty pounds of duct tape and cardboard and entered it in the Cardboard Classic. There were no winners or losers then. The Heartbreak Hotel went through the hay bales straight to the side of the Thunderhead Lodge.*
>
> *My wife doesn't like to go up there anymore because she remembers the way it was and how pristine it was. They did a nice job* [with the development], *but things change.*

The springs, initially due to their location outside town, were fairly protected from overuse, but in the 1970s, as the town expanded, people came to live near the springs, leaving all sorts of garbage strewn around. Some brought and left mattresses that they slept on. Although the hot springs were owned by the SSHRA, it didn't have the funding to develop or care for the Strawberry Park Hot Springs as well as the Old Town Hot Springs. Nor did it have the manpower to patrol the area. For these reasons, the SSHRA started looking for a buyer in the late 1970s. Add to that the unchecked nudity, and some members of the community were upset.

Nude bathers were scaring off some families and children from using the area. The Routt County sheriff attempted to stifle the nude bathing, as recounted in the *Hayden Valley Press* in May 1977:

> *Nudity a no-no. Hot Springs getting "hotter"*
>
> *Routt County Sheriff's deputies under the direction of Sheriff J.T. Kelton, have cracked down on nude bathers at the Hot Springs, a well-known area to most local residents. Sheriff's deputies have issued 23 summons within the last three weeks and expect to increase the patrols in the scenic area as the long, hot summer draws on.*
>
> *With the start of the summer tourist season, deputies merely warned those bathing in the nude that they were in violation of Colorado law and further occurrences would result in the issuance of citations and appearances in County Court. After encountering repeated offenders and receiving numerous complaints from local citizens and tourists, deputies began issuing citations. Those cited into County Court are charged with indecent exposure and face a penalty from $50 to $750 fine and/or 6 months in jail. Indecent exposure is the charge recommended by the Office of the District Attorney for this area.*
>
> *Sheriff Kelton advised he has received a letter from the Steamboat Springs Health and Recreation Association, which owns the Hot Springs area, asking that his department patrol the area and enforce the law pertaining to indecent exposure. Sheriff Kelton further stated that his deputies had also escorted several classes of school students to the Hot Springs for school-sponsored picnics.*
>
> *"We have received several comments from school personnel and mothers about how nice it was to see children enjoying the area once again," says Sheriff Kelton. "It actually requires a small effort on the part of bathers to wear shorts or swimming trunks and allow everyone the use of the natural warm waters."*
>
> *"We're issuing summons and complaints in an attempt to expedite the judicial process," Kelton said, "But if this method doesn't prove satisfactory, we'll take the necessary steps to tighten up even more."*
>
> *"We're not launching a personal vendetta against any particular group or faction," Kelton continued, "But by the same token, the Hot Springs is not going to become someone's private swimming pool and we intend to keep it available and respectful for the general public," he concluded.*

Despite the sheriff's best efforts to make the place reputable again, nudity prevailed.

One Steamboat local recalls visiting the springs around that time with her husband and her very conservative in-laws. She held her breath as a nude man approached her in-laws. The man apologized, saying, "I hope I didn't scare you." The woman's mother-in-law looked at the nude man and said, "Honey, I've seen it all."

However, the SSHRA felt the springs would be better off privately owned. In 1982, Chicago-based Don Johnson purchased the springs. He developed and cleaned them up and regulated them so that locals, families and tourists could all enjoy the area. Today, there are three pools divided by hand-built stone walls with five waterfalls around the area. The creek is used to regulate the temperature from one pool to the next. Depending on the time of year and temperatures, the largest of the pools is the hottest at about 104 degrees.

There is a changing room and bathroom above the springs for visitors, as well as a tepee and a spring-heated stone changing room closer to the springs. However, because the springs are in a remote location, without lighting at night, things can still get a little wild up there.

As Rodney Pryde recalls, "I remember rolling up there with a crew of hot, young ski bums and bunnies, getting nekkid in a teepee, sprinting to the hot pools, only to find a busload of equally nekkid seniors doing their thing."

Teenagers would go to elaborate lengths to sneak in. Employees working at the entrance of the springs count cars to make sure there weren't more cars than people. As a teenager, Chris Erickson and her friends would outsmart the system:

> *Everyone would chip in a dollar for whoever was driving the largest vehicle that night, and you would pack as many people into that as vehicle as possible, which meant ten, fifteen people into one vehicle because they used to count cars. Then the driver would drop everyone off at the saddle* [before the springs] *and we'd sneak all the way around the back and get in. We had to be cautious. We couldn't all leave together because the old guy worked there. Les, he was there forever and ever, and he was on it.* [After you soak] *you leave in groups of ones or twos. The first people out would have to wait for the entire group to come out in the middle of the winter, which could be ten minutes or more in subzero temperatures.*

Les was a staple at the hot springs for a long time. Many people thought he owned the springs. He was a tough man, to say the least. When you saw

The changing room at the Strawberry Park Hot Springs is heated with steam from the springs. © *Dagny McKinley 2012.*

him come down with his flashlight, you knew it was time to get out. He didn't put up with nonsense from anybody and was as much a fixture of the hot springs as the springs themselves. While most people have a story of being yelled at by Les, Suzie Hawkins saw the softer side of this man. Suzie used to own Iron Horse Inn:

> *We used to always take customers to the hot springs, and I would purchase an annual pass. You could always bring a guest with the pass. I went often. The hot springs have always been a special place to me, but it's hard to articulate what made it so special. One night a friend called and said, "I'm coming over from Winter Park and wondered if you wanted to go to the hot springs." I was all over that idea. This was back in the days when the springs were able to stay open much later because I never closed my desk before 10:00 p.m.*
>
> *At the hotel, if a guest ever paid for a room in cash, I would always take the money out of the drawer because it was always too much of a temptation to anybody who worked for me. Keeps everybody honest. Somebody had paid in cash, and there was a $100 bill in there, and I saw the money and stuck*

> *it into my bra, which was sort of standard procedure for me. That evening my friend picks me up and we go off to the hot springs. We get up there and rip our clothes off, and everything was absolutely fabulous.*
>
> *The next day, I'm sitting in church and thinking about that $100 bill I was going to put in the collection plate and realized I lost it at the springs last night. I was so mad at myself. I came back home and my friend was still there, and he said, as men will do, "Well, why don't you call your friend Les? You seem to be the only person in the world that he's ever nice to, and ask him if he found it." So I said, "I think I will." So I called my friend Les, who was the ultimate curmudgeon, and I said, "This is Suzie from the Iron Horse Inn." "Oh, hi Suzie, how are you?" "I'm fine, thank you, Les, for asking. I was up there last evening." "Oh yes, I saw you." "And I lost $100 bill." "Oh, was that yours Suzie? I wondered who that belonged to. Will you be up again soon or would you prefer I bring it to you?" Of course I said, "If you wouldn't mind, it would be great if you would bring it to me. I don't have time to get up there." And he brought it to me.*

There is a paradox at the springs. From a family recreation area to a party place, from day to night, everything seems to shift.

Drinking and partying were common for a long time at the springs until one night ended in tragedy. In March 2003, a woman was found dead in the springs. She had drowned. Despite the best effort of nurses on the scene administering CPR and emergency responders continuing the resuscitation, the woman could not be saved. Patricia Maestas was at the springs with her husband, reported the *Steamboat Pilot*. The article goes on to say that her husband had left to get a towel, dry off and get dressed. When he returned, he couldn't find her. He yelled for her, but there was no response.

Unfortunately, toxicology reports showed Maestas had not only alcohol in her system but cocaine and methamphetamines as well. Police believe these were the major causes behind her drowning. Her blood alcohol levels were .288. Police believed she passed out and drowned. There were no indications of a struggle or physical violence. The 3.2-percent-alcohol beer that was once permitted at the springs was no longer allowed. Although the springs used to close to new visitors by 10:30 p.m. and everyone had to be out by midnight, the enforcement on those hours was a little lax.

Drownings are one of the leading causes of death in national parks and wildlife recreation areas. There are no lights at the springs so people are advised to keep tabs on friends and proceed cautiously.

Because the hot springs are outside of town, they were able to abide by different rules and constraints than those in town. However, in 2003, the city took Johnson to task for the late hours being kept, while at the same time the townspeople were petitioning for later hours.

The hope for many was that the springs could stay open on weekdays until midnight. Steamboat's community is based on conservative beliefs, and the hours were denied. Tensions rose because people using the springs sometimes forgot to respect the privacy of the neighbors who live in the area. According to one neighbor, Butch Bowers, he "has been woken in the middle of the night by people camping in his yard or pounding on his door. "They either want to use your phone or come in your house to get warm," Bowers said in an interview with the *Steamboat Pilot*.

The city showed flyers with the hours already claiming the springs were open until midnight and berated Johnson for ignoring ordinances already in place.

Despite the debauchery that does or doesn't take place at night at the springs, there is still a community feel to an area that offers the combination of recreation and wilderness that many moved to Steamboat for. Sometimes the Boulder Hot Springs Team (one person) will show up to challenge the Steamboat Hot Springs Team (one person). The challenge is to see who can last the longest in the springs without getting out. Upward of three hours sometimes pass before, skin pruned and body dehydrated, the two teams shake on a draw.

In spring, runners can look forward to the Hot Springs Shortcut race, where people have to find the fastest route from downtown to the springs. There are no rules for where you can and can't go. People are encouraged to jump fences and cut through backyards. The only danger is getting caught in the brush, which can slow you down. The race has been taking place for more than twenty-five years. On May 12, 2005, before the twenty-year anniversary, the *Steamboat Pilot* covered the event:

> *It's part of the 20th annual Hot Springs Shortcut race, an event set aside for those who navigate their way from downtown Steamboat Springs to the Strawberry Park Hot Springs in the fastest time possible.*
>
> *"One year I went out with my stop watch before the race and tried this route and that route," said third-year race director Bill Philip. "A few years ago, there were some cross-country coaches who got permission from property owners to go over fences."*
>
> *Several people swear they have the fastest route. On Saturday, the participants will find out who is right.*

> *The Hot Springs Shortcut is the second race in the Steamboat Springs Running Series, and it's open to walkers, runners and mountain bikers.*
>
> *Mountain bikers must follow a set route from the Steamboat Springs Health and Recreation Center to the Strawberry Park Hot Springs. Walkers and runners can go any way they wish, providing they don't trespass on property where they are not wanted.*
>
> *Typically the race is held on Memorial Day weekend…When runners and walkers filter out at Strawberry Park Elementary School, they will follow Routt County Road 36 past The Lowell Whiteman School to the dirt road that climbs to the Strawberry Park Hot Springs.*

The fastest route may, in fact, be the route that the water is believed to flow underground from the Strawberry Park Hot Springs to the Old Town Hot Springs. The race is the only one that allows runners, walkers and bikers to compete on the same course at the same time. Winners and losers are given free admittance to the Strawberry Park Hot Springs to soak away the aches of the day.

The hot springs are more than just a place where people visit and soak. Body and water massage are offered to lengthen the body and rehabilitate injuries. In 2005, the springs offered Waterdance massage. The experience was one of a kind, as Autumn Phillips reported in her article "Returning to the Womb" for the *Steamboat Pilot*:

> *The first step in the Water-dance is trust.*
>
> *The Waterdance is a new kind of massage being offered at the Strawberry Park Hot Springs that takes place almost completely underwater.*
>
> *And before massage therapist Michael Moss can even begin, he must break through all your fears—fear of being held underwater, fear of drowning, and fear of giving up control to another person.*
>
> *The Waterdance sessions are an hour long, and for first-timers, Moss will spend as much as 15 to 30 minutes just getting them comfortable.*
>
> *"They need to feel safe," he said.*
>
> *The water is heated to 96 degrees, skin temperature. The massage begins on the surface of the water where Moss loosens your joints and your breathing by moving you back and forth like river grass in a current.*
>
> *Your eyes are closed. The sun shines down on you. Your ears are underwater, and all you can hear is your own breathing. You feel like you are being rescued. You are dying. You are being born. The water and the movement of your body in the water removes you from the stress of the day and transports you to a different place.*

"For some people, it's a roller coaster ride," Moss said. "For others, it's more like being in the womb."

As Moss moves your body, stretching muscles and working out tension, he matches his breath to the rhythm of your own.

"I breathe with you," Moss said. "As you inhale, I inhale. As you exhale, I exhale. That simple act is very nurturing." Your breath begins to lengthen, and Moss prepares you to go underwater.

You inhale and Moss pushes you under the water. At this point, you might imagine that you would panic and fight for the surface.

Instead, you are relaxed. The preparation worked.

"A diver's reflex kicks in," Moss said. "When you're skin diving, the body just takes care of the core. You can stay under for 30 seconds, 40 seconds, even a minute and a half. You enter a meditative state."

The first time Moss gives a Waterdance massage, he intentionally keeps it on the conservative side. He takes you underwater, stretches you or works on a knot in your muscle and then brings you back up for a breath.

"Because I breathe with them, I have a sense of when they need to breathe," he said. "I try to make the time below the water and above the water seamless. The shift is invisible."

Waterdance is similar in many ways to Watsu, a form of aquatic massage offered at Strawberry Park Hot Springs for the past eight years. The difference is that Watsu takes place completely on the surface of the water.

Arjana Brunschwiler and Aman Schroter created Waterdance in 1987. The moves are inspired by aikido, dolphin and snake movements, and dance—all synchronized with the breath of the practitioner and the client.

Without the constraints of the massage table or even the constraint of keeping the body above water, the therapist can move your body in any direction.

"On a table, I can work on your back or on your front," Moss said. "Waterdance is dynamic. I can work in all planes at once."

In June 2005, another death plagued the Strawberry Park Hot Springs when a man in his mid-thirties was found unconscious in the upper pool of the springs. The man was found in the early evening and rushed to Yampa Valley Medical Center. He arrived alive but died at the University of Colorado Hospital in Denver. He had been at the springs alone. Five people with CPR certification took turns resuscitating him. Causes were believed to be natural.

Despite the few tragedies that had occurred, by December of that same year, the contribution of the springs to the history of Steamboat became apparent when the Strawberry Park Hot Springs was placed on the Routt County Register of Historic Places by the Historic Preservation Board. The honor is for the springs and the surrounding physical area but not for the buildings constructed by Johnson in the 1980s, as the construction is considered too recent to be historic. The buildings have been constructed with a rustic look that blends well into the landscape.

Throughout the years, different events have come and gone at the hot springs, but a favorite that held the imaginations of many was the fire dances. Fire throwers and dancers came up from Denver with friends. They would hook up with Steamboat's African Drum Troupe and put on a show under the full moon.

Allison Plean recounts a night at the hot springs with the fire dancers in her article "Playing with Fire" for the *Steamboat Pilot*:

> *Danger is just part of the experience—and allure—for fire dancers and poi spinners.*
>
> *"You get used to losing eyelashes and nose hairs and routinely losing arm hair and eyebrows," fire dancer/poi spinner Erica Olson said. "After spinning for a year, people get cocky and that's when accidents happen. Newbies don't burn themselves because they are overly cautious. When you do, though, it's a mark of pride."*
>
> *Olson, who has been spinning fire for five years, and Chloe Marcellus, who has been spinning fire for three years, will perform tonight at Strawberry Park Hot Springs.*
>
> *The best seat for viewing is in the hot pool.*
>
> *"Most people sit in the hot pool on flagstones with the steam coming up around you," Olson said. "You can't see the performers. All you can see is these loops and whirls of lights."*
>
> *The Steamboat African Drum Troupe will provide the beat for the fire dancers, who use various instruments to "hold" their fire.*
>
> *"People dance with poi, fire fans, nunchucks, fire fingers, palm torches, fire hoops, fire robes and headdresses," Olson said. "If it can be set on fire safely, someone will do it."*
>
> *Olson spun "dry" for six months before she lit her poi. Fire dancing, she said, involves safety and skill.*
>
> *"It requires coordination, dance, agility, grace and interacting with one of the most dangerous of natural elements in a very intimate way," Marcellus said. "The first time I saw it, I thought it was the hottest, most unconventional form of eroticism I've ever seen."*

Olson and Marcellus insist they are not pyromaniacs.

"Though I would search for things to blow up when I was little," Marcellus said. "Which is something your parents taught you not to do—play with fire."

Fire spinning has been growing in popularity. In the past six or seven years, the number of fire spinners in Colorado has increased from just a few to a couple hundred, Olson said.

The element of fire has existed as a part of tribal dancing for ages.

"Poi developed in different places around the world," Olson said. "The Maori people of Australia and New Zealand used it for weapons training, and the women used it for a dexterity exercise."

There are at least a handful of spinners in Steamboat Springs.

"For a lot of people, it's a very spiritual thing," Olson said. "The rush is really being able to expose people to a whole new world that they never dreamed of."

The beauty and uniqueness of the area attracts those longing for peace and beauty in nature, but the good isn't always what's attracted to out-of-the-way places. Nudity, death and then robbery. There is crime in Steamboat but generally not enough to make people nervous outside at night. In 2011, there seemed to be a rash of violence and crime. On January 22, 2011, a white man put on a ski mask and a dark-colored hip-length coat with yellow and brown camouflage pants and gloves. When he arrived at the springs, he brandished a gun to get $800 cash from the employee on duty at the springs. With money in hand, the assailant pepper sprayed the employee and made sure the phone was disabled.

Incidents like that are rare, and in remote places without cell phone coverage, people can get away with more than in town. Drive time in bad weather can be anywhere from twenty minutes to a half hour, leaving those out of reach out of luck until someone can get to them.

As mentioned previously, everyone in town seems to have a story about these springs. There is a draw to the place and a calm energy that surrounds the area. The hot springs hiking trail attracts those who want a little exercise before a soak, those looking for a quiet jogging trail or those who simply need a little shelter from life.

Many of the stories told about the springs are fun, but there is one story—a love story—that should also be told. It involves the balance of emotion and logic and how two near strangers' lives were transformed by a single encounter on the hot springs trail.

A squirrel perches on a tree on the hot springs hiking trail. © *Dagny McKinley 2012.*

The story begins with two lives being lived as lives often are—following passions, struggling for dreams. Kim is an artist in town, best known for her bodyscape photography, with which she turns women's bodies into black-and-white landscapes. Kim is constantly pushing her art and the forms it takes to new levels. In 2012, she constructed a cocoon from papier-mâché, words from books, rose petals and soul for a performance art piece held at Harwigs L'apogee, a local restaurant. The cocoon was a vision she had for a long time, yet she wasn't sure what the story of the cocoon should be. The performance was successful, with a half-naked nymph in glitter emerging from the cocoon, but this still was not the story Kim wanted to tell.

The day after the performance, Kim unhooked the cocoon and transported it to the deck of her home, where it sat. And sat. Through leaves falling from trees and snow falling from clouds, the cocoon waited patiently for a home. Raccoons found shelter in the warm depths and also a meal or two. Some of the papier-mâché crumbled. Pieces deteriorated. And then a new life appeared. This life had already seen deterioration and was rebuilding not a cocoon to hide away from life but a nest that good things could fall into.

Jay's charitable contributions in town always get people talking. His mind is logical yet futuristic, as he can see where technology is heading and where opportunities can be had. He is almost too far ahead of his time for a small town but not for a small-town artist whose work could stand in any city gallery. One night, fate brought the two together. Walls were broken down, conversations and kisses exchanged. From that night, something grew. The nest was being lined with warmth.

Butterflies cannot fly unless they leave their cocoons behind.

The greatest tree in Colorado stands along the pathway of the Hot Springs trail. The branches curve and arch, offering shelter and respite on a sunny day. Once, animal skulls hung from the branches, but it was time, maybe, for death to be replaced with a dream. Jay and Kim chose one of the coldest nights of the year, a night when memories could fall to the ground like snow and be frozen to the earth. They trudged down the trail, the cocoon on a sled, their hands numb, until they reached the tree. The hanging process did not take very long. The cocoon was illuminated once, captured by camera and then left to the wild. The couple did not waste much time in the cold. They returned up the trail, the cocoon of their former lives hanging in a tree, sheltering the past and giving them freedom to move toward the future and toward a deeper love.

Love is why the springs are still visited regularly by many. There is a special feeling in the air, and there is still the edge of wildness, the possibility of the encounter with a bear or a moose, crow or hawk. Today, the springs face a new threat, but not from people. Drought over the previous year brought water levels down and came close to closing the springs for part of the season. Water rights are a huge issue in a town where ranching, recreation and necessity all fight for survival. These springs reflect nature's patterns and offer a place where one can contemplate life, beauty and the pure joy of life.

Chapter 12

Sulphur Cave Spring

Mineralogy of Spring, 2012

Temperature	61 degrees
Alkalinity, mg/l	1340
Bicarbonate, mg/l	1340
Boron	none
Calcium, dissolved, mg/l	29
Carbonate, mg/l	none
Chloride, mg/l	1300
Coliforms, -/100 ml (1/100 ml)	none
Fluoride, mg/l (4 mg/l)	3.2
Hardness, mg/l	204
Magnesium, dissolved, mg/l	32
Nitrate, dissolved, mg/l	none
Nitrate/Nitrate, mg/l	none
Nitrite, dissolved, mg/l	.01
Nitrogen, mg/l	none
Ph, units	8.1
Potassium, dissolved, mg/l (5,600 mg/l)	120
Sodium absorption ration	68.44
Sodium, dissolved, mg/l (3,000 mg/l)	2220

Silica, mg/l	none
Sulfate, mg/l (250 mg/l)	836
Arsenic, dissolved, mg/l (0.05 mg/l)	.099
Barium, dissolved, mg/l (1.0 mg/l)	.04
Cadmium, dissolved, mg/ (0.010 mg/l)	none
Iron, dissolved, mg/1 (0.3 mg/l)	.03
Lithium, dissolved, mg/l	3.45

mg/l = milligrams per liter
Elements: Maximum recommended by state and federal government in ()
*Source: Tread of Pioneers Museum, "The Springs of Steamboat Springs Walking Tour"

LOCATION

The Sulphur Cave Spring can be accessed two ways. The first is from the base of Howelsen Hill. There is a parking lot toward the baseball fields and a road that leads up to the Sulphur Cave Spring. In the winter, this area is covered in snow and can easily be skied to.

The other way to access the cave is to follow Lincoln Avenue (Highway 40) through town to 13th Street. Turn left here. Follow the road over the train tracks and take a left at Gilpin Street. At the stop sign, turn left again onto Fairview Avenue. Take your first right. At the top of the street, you'll see parking and a sign for the Sulphur Cave. The walk takes you up a small hill and then dips down toward Howelsen Hill. The Sulphur Cave Spring is blocked off by a wooden fence not only to warn of the dangerous gases that are emitted but also to prevent any skiing or boarding accidents in the winter when the cave can become buried under snow.

WHAT TO EXPECT

The cave today looks much as it did over one hundred years ago. Because of its location out of the way, not many people know it exists. Before you reach the spring, you will be greeted with the strong smell of sulphur. The cave is limestone, and a narrow opening allows a glimpse of the waters running down through the cave. Follow the trail of water up the hill to see where the

The interior of the Sulphur Cave Spring. Water flows through cracks and crevices. © *Dagny McKinley 2012.*

spring comes out of the hill and trickles down in a slow-running stream into the cave. A flashlight or spotlight helps to get a better view inside, as the sun never quite seems to reach the perfect angle to illuminate the interior. When Steamboat was first settled, people used to picnic on top of the cave. One of the first cabins ever built in Steamboat, by trappers, was just above the Sulphur Cave. The cabin was later destroyed by an avalanche.

This is one of those places that made me curious to explore because people used to have visions of the future after inhaling sulphur fumes. I was also curious because very few people have ever been in this cave. The fumes are enough to kill a person, and without the right equipment, a human couldn't survive more than a very short while inside.

History

The Sulphur Cave Spring has an aura of mystery about it. The entrance is a cavern in limestone that offers glimpses of the water-carved rock formations within. The cave is filled with a highly toxic hydrogen sulfide gas, carbon monoxide and carbon dioxide. In ancient days, oracles used sulphur vapors and caves like this one to prophesy future events. They bravely entered these caves and inhaled the fumes until visions came. A few breaths inside the Sulphur Cave Spring could induce unconsciousness. Legends of Steamboat claim the cave was used to torture enemies through asphyxiation. There are stories that people have died in this cave in more modern times, but no records can be found to say whether that is true. Someone with curiosity and without knowledge could easily be enticed to venture below ground to see what unique creatures dwell there.

While rumors of danger and evil abound, there are also more heartwarming tales about this spring. Some say the Ute Indians who lived in this valley set up a sign to warn newcomers of the dangers. Others say settlers hid from the Indians here during tensions, despite most accounts that show the Utes did not ever come to Steamboat with hostile intent. Given what is known of this cave today, it seems highly unlikely anyone would have been able to stay in the six-foot crawl space in the cave for long with such toxic vapors. American Indians believed that the mineral springs were filled with spirits. The smells of sulphur and bubbling waters were proof of this.

The fascination with this spring is evident through an early report of the area from the collection of the Steamboat Springs High School Library:

The Cave

Surrounded by several of the important springs of this group is a cave of singular interest. The floor of the cavern has an incline of about 10 degrees or 15 degrees from a horizontal. From this mysterious abyss comes forth gases and vapors that are fatal to life until diluted with the air. Standing at the mouth you can look into a point where suffocation and death would ensue. The Indian regarded this as the threshold to the lower regions, the abiding place of an enraged evil spirit, whose only mission was to breathe torture and death to all who entered. From the trapper, missionary and Indian have been gathered many tales connected with this spot. One whose authenticity is unquestioned is that it was the place of execution of the Ute Indian who was condemned to death by the grand council of his tribe. Upon confirmation of sentence by the head chief, the condemned, guarded by six armed warriors, each having his right hand painted blood red, was conducted to the mouth of this breathing monster, where the palms of his hands were placed together over his head, and his wrists bound by a cord some forty feet long which was made by four strands of the sinews of a young bison braided together. At the word of command he began quickly to descend. If he refused, or faltered, each guard shot into his body a poisoned arrow which caused death in the utmost agony.

Today, while no more rituals take place around the Sulphur Cave, as far as is known, and it is fenced off to indicate the potential danger of entering the cave, the mysteries of the Sulphur Cave Spring are just beginning to be revealed.

Another more scientific description written in 1939 by Thomas J. Farnham in his essay "From Travels in the Great Western Prairies, the Anahuac and Rocky Mountains, and in the Oregon Territory, August 4th, 1839" recalls:

We were early on route this morning, down the banks of Little Bear River; course north-west. Our track lay so low, that the mountains were seldom seen. A portion of the Anahuac ridge in the southwest, was the only height constantly in view. The plains, as they are called, on either side of the river, were cut into vast ravines and bluffs. In their side sometimes appeared a thin stratum of slate. Few other rocky strata were seen during a march of fifteen miles. About twelve o'clock, we came upon a cave formed by the limestone and sulphur deposit of a small stream that burst from a hill hard by. The water had, by constant depositions, formed an elevated channel some five rods down the face of the hillside, at the termination of which it spread itself over a

Folks visiting the Sulphur Cave. They are thought to be friends or family of Helen Gooding Reader. *Courtesy Tread of Pioneers Museum.*

> *circular surface of one hundred and fifty or two hundred feet in circumference. In the centre of this, was an orifice, down which the water trickled into the cave below. As little of the cave could be seen from the ground above, myself and two others attempted to explore it. We found the roof hung with beautifully crystallized sulphur, and the bottom strewn with large quantities of the same material in a pulverized state. The odour was so offensive, however, that we were glad to retreat before we had formed a very perfect estimate of its extent and contents. It was about six rods long, eight feet wide, and four feet high. Near it were a number of warm springs. On the bluff, a few rods above it, was a small tract of fused rocks. In all the circle of vision, however, there were no elevations that indicate any powerful volcanic action in former times; nor any from which these rocks could have tumbled or been thrown. The warm springs, however, in the vicinity may, perhaps, indicate their origin.*

Children were often warned to stay away from the spring. Tom Wither recalls being forbidden to go anywhere near the cave. In the summer, he and

Sled run at Sulphur Spring. Sulphur Cave Spring sits in the shadow of Howelsen Hill, the first ski-jumping hill in Steamboat. *Courtesy Tread of Pioneers Museum.*

his friends would tell their parents they were going to play near Howelsen Hill. Their parents would warn them not to go anywhere near the cave, so of course Tom and his friends would peck in as far as they felt safe and then would scare one another and run out.

After skiing came to town, the cave was included in recreational activities such as sledding. However, the interest as to what might be found in the cave persisted.

Cavers aren't the only ones curious about the depths of this fairly small cave. Because caves are places of relative isolation and because of the toxic gases of the sulphur cave, insects and formations inside may be unique to Steamboat. So when Fred Luiszer, geologist and speleologist with the University of Colorado in Boulder, decided to go in to explore, he had high hopes that weren't disappointed. Several articles in the *Steamboat Pilot & Today* detail his adventure in the cave.

In order to ensure the safety of those going into the cave in 2007, Luiszer and his crew brought equipment that pumped out the toxic hydrogen sulfide gas. "Many personal safety gas detectors are set to alarm when hydrogen sulfide reaches 10 parts per million, and most employers won't let workers in an environment for long periods of time at 15 parts per million," said Luiszer as his alarm started to beep, indicating the air was at 35 parts per

This image shows the proximity of Sulphur Cave Spring (fenced) to Howelsen (right) and Mount Werner (ahead), showcasing the two main interests that caused Steamboat Springs to be settled: mineral springs and outdoor recreation. © *Dagny McKinley 2012.*

million. "It's time to get those guys out of there," he said. Luiszer added that the gas doesn't pose long-term risks unless the parts per million reach 150, when olfactory nerves are paralyzed after a few inhalations.

No one knows what effect that temporary disruption of environment, as well as the intrusion of humans, had on the inhabitants of the cave; however, Luiszer considered the expedition to be a success. The first exploration revealed rare gypsum crystals. "Gypsum crystals form when bacteria metabolizes the hydrogen sulfide gas and interacts with the limestone cave walls," said Luiszer. But the biggest find of the day was snottites, a bacterial colony, or formation, similar to stalactites that resemble human snot. They look slimy (think of snot running out of a nose) and sway with air or wind. Scientists now believe that hydrogen sulfide may be a critical factor in the formation of snottites. They are currently only known to exist in one other cave in Mexico. Luiszer hopes to come back with more funding in order to explore the cave further.

Years may pass before we fully understand the extent of the delicate environments of caves and springs. In the future, we may realize healing benefits that we are unaware of now or unique life systems that can only thrive in Steamboat's unique environment. Hopefully, human exploration will not cause the disappearance of any species before we are able to understand their purpose.

"It's a unique cave that needs to be preserved," Luiszer said. "It's a scientific treasure." Beyond that, Sulphur Cave Spring is one of Steamboat's treasures.

While the Sulphur Cave Spring is now fenced off, the area cannot be ignored by those who spend time on Howelsen Hill. This cave's location is the epitome of the two reasons this town has made a name for itself: the mineral springs and skiing.

Chapter 13

Sulphur Spring

Mineralogy of Spring, Early 1900s

Bubbling Spring Temperature 75 degrees F

	Parts in 100,000
Sulphate of Lime	49.81
Sulphate Potassa	27.30
Sulphate Soda	16.11
Chloride Magnesia	14.50
Chloride Sodium	193.13
Bicarbouate Soda	423.41
Bicarbonate Iron	0.39
Bicarbonate Magnesia	0.35
Total Solids	725.00

Mineralogy of Spring, 2012

Temperature	73 degrees
Alkalinity, mg/l	2390
Bicarbonate, mg/l	2390

Boron	none
Calcium, dissolved, mg/l	35
Carbonate, mg/l	none
Chloride, mg/l	1250
Coliforms, -/100 ml (1/100 ml)	none
Fluoride, mg/l (4 mg/l)	3.4
Hardness, mg/l	235
Magnesium, dissolved, mg/l	36
Nitrate, dissolved, mg/l	none
Nitrate/Nitrate, mg/l	none
Nitrite, dissolved, mg/l	none
Nitrogen, mg/l	none
Ph, units	7.8
Potassium, dissolved, mg/l (5,600 mg/l)	120
Sodium absorption ration	57.68
Sodium, dissolved, mg/l (3,000 mg/l)	2010
Silica, mg/l	none
Sulfate, mg/l (250 mg/l)	584
Arsenic, dissolved, mg/l (0.05 mg/l)	.087
Barium, dissolved, mg/l (1.0 mg/l)	.02
Cadmium, dissolved, mg/ (0.010 mg/l)	.006
Iron, dissolved, mg/1 (0.3 mg/l)	none
Lithium, dissolved, mg/l	3.35

mg/l = milligrams per liter
Elements: Maximum recommended by state and federal government in ()
*Source: Tread of Pioneers Museum, "The Springs of Steamboat Springs Walking Tour"

LOCATION

West Lincoln Park on the corner of 13th and Lincoln.

What to Expect

It's amazing how odiferous this spring is. You can tell which way the wind is blowing depending on where you can smell the spring. On a good day, the sulphur smell travels down to 9th Street and beyond. Driving into town, visitors start to look at one another to see who is responsible before they realize it's sulphur gas and not any other kind of gas. In fact, the strong smell that wafts over town earned the park the nickname "Fart Park."

This is a cool spring because of how active the waters are. They fizzle and bubble as I imagine a witch's cauldron would. Children and adults are still mesmerized by the waters. These waters, of all the springs, feel like they have life boiling up out of them, like they are still vigorous and young and alive.

History

The Ute Indians believed the Sulphur Spring, also known as the Bubbling Spring, had curative powers. They used to bathe their horses in these waters before battle. Prior to development, deer, elk, horses and other large animals were attracted to the salty grasses around this spring and often drank from the sulphur waters.

The sulphur spring was the popular bathing place for visitors and settlers until the Old Town Hot Springs were discovered. The waters are lukewarm but warmer than the river and less of a hassle than boiling water for a tub on the kitchen floor. Even a lukewarm soak while living and surviving in such rough terrain had more than physical relief; it also offered mental relief, as noted in the following from the collection of the State Historical and Natural History Society:

> *The Bubbling Spring*
> *The Bubbling Spring was once called a most valuable spring with many cures to its credit. This spring was 15 feet in diameter and several feet in depth. The surface rolls and tumbles in constant and violent ebullition, hissing and crackling as though a great furnace were somewhere beneath. The observer is impressed and fascinated by the strange spectacle, lingers by and returns day after day to view its incessant turbulence. The motive power of all this commotion is not heat, but comes from far down in the earth's primitive depths; is the escape, under high pressure of great volumes of*

Sulphur Spring and Cabin Hotel in the background, 1930s. *Courtesy Tread of Pioneers Museum.*

> *carbonic acid, gas, sulphuretted hydrogen and various other gases. At times the pungency of these gas odors is so great that the bather finds it necessary to step for a moment from their midst into the natural air. Aside from the analysis, which indicates a bicarbonate water without a rival, this spring has already established its therapeutic value. It is the idol of its patients. Especially do the rheumatics love to stand by and tell how they came here, crippled, suffering and reduced to the last stages of helplessness, point to their now supple joints, free from pain, enjoying robust general health and declare "that spring did it," never losing an opportunity to pour into all ears that will listen, a song of its marvelous virtues. Even the Ute Indians' laconic reference to it is, "Sta Wano," (good very). Gout, rheumatism, scrofula, cutaneous diseases and blood poison here find relief and cure. The effect of sulphuretted and carbureted hydrogen gases is to powerfully excite function of the skin and to very quickly restore it to a normal condition.*

The Crawfords, the first family to settle Steamboat Springs, often bathed in the Sulphur Spring before the Bath Spring was discovered. Today, people are still mesmerized by the bubbling turmoil that lives in this spring.

When driving into town, if anyone is searching for a word to describe the nasal experience they are having, keep in mind this spring has politely been called "fragrant" by some. For brave souls, there are no fences to prevent a quick dip in the waters, but waiting until the snows melt is highly recommended, as the spring won't do much to melt away cold air temperatures.

Chapter 14

IN MEMORIAM OF LOST SPRINGS

"Civilization does a lot to spoil natural beauties," said Lulita Crawford Pritchett, granddaughter of the founding family of Steamboat, in reflection of the Navajo Spring. Of the 150 springs that once populated the area, only a fraction still exist that can be enjoyed by the public. Once-famous springs such as the Navajo Spring, Gazebo Spring, Ouray Spring, Juanita Spring and No Name Spring simply no longer exist, at least above ground. According to Pritchett, there were other smaller springs to the west of the Lithia Spring that were trampled by logging operations. Bird's Nest, Fairy Lake, Frying Pan, Doctor, Aeolian, Galvanic, Annie, Baby, Hades, Magnesia and Moffat are just a few of the other springs mentioned in the *Steamboat Pilot* on June 16, 1988, as having disappeared. The following is a dedication to the springs that no longer have a voice or a presence in this town.

The Chinking Spring is one of the springs that sits underground due to development of the city. The mud that was once found around this spring was used as a sealant that held the logs together on the cabins of the early settlers of Steamboat.

The Navajo Spring was once located near the cabin of the second settlers in Steamboat who homesteaded the area in 1876, according to the *Steamboat Pilot* in March 1976. While the Navajo waters were cold and sparkling, the sulphur within the waters allowed the spring to be overshadowed by the

Soda Spring. On November 8, 1967, Lulita Crawford Pritchett wrote in a letter to a Mr. Max H. Selbe:

> *I think Grandfather's brother John originally took up the property where the Navajo Spring is located. I'm sure it was not included in the Steamboat Springs town site. Later, I think Grandfather's son John (One-time county clerk) owned the place.*
>
> *There used to be a nice little wooden pagoda over the spring, but of late years that has been removed and no care has been taken of the spring.*

The Navajo Spring Mineralogy:

	Grains in One Gallon
Calcium Sulphate	16.28
Sodium Sulphate	21.32
Iron Carbonate	14.20
Potassium Carbonate	12.16
Silica	.30
Alumina	.54
Carbonic Acid Gas	Trace
Total Solids	68.80

No-Name Spring: The *Steamboat Pilot* reported on Thursday, June 18, 1988:

> *City officials and others [recently] thought this was the Navajo Spring, but further research confirms that Navajo Spring is in a vacant lot on the west side of the Fairview subdivision. It was covered with a manhole by a former owner and is now pumped into a runoff stream nearby. Two other smaller springs, Juanita and Ouray, were once near the Navajo Spring as well.*

Conclusion

History is determined by the stories we choose to tell. The importance of a place is easily lost when we don't know or we forget the roots of the area and the stories that were told before we arrived. The waters that flow underneath Steamboat Springs have helped heal people, relaxed people, provided for recreational activities and mental balance and tickled more than a few taste buds. The histories and stories in this book have been written and preserved in the hopes that the mineral springs will always be remembered and celebrated.

Progress is measured in the way we respect our legacy. Balancing a growing population with preservation can be a difficult task. However, the one thing we know for certain is that there are bound to be mysteries discovered within the springs and caves of Steamboat. These discoveries may not be realized for decades or centuries to come. Only in the future may we truly understand the treasures these springs offer to our community, our environment and the creatures that live in, on or off the springs.

Bibliography

Crawford, James Logan, ed. *Steamboat Springs and Its Founder: An Interview with James H. Crawford.* Recorded May 24, 1923, by Thomas F. Dawson of the Colorado State Historical Society.

———. "Steamboat Springs & Routt County Colorado brochure." Compiled and published by the Northwestern Loan & Investment Company Steamboat Springs, Colorado, *Pilot Print*, Steamboat, circa 1910.

Crawford, Lulie Margie. Contributor Lulita Crawford Pritchett. *Diary of Lulie Margie Crawford: A Little Girl's View of Life in the Old West, 1880–1881.* Steamboat Springs, CO: L.C. Pritchett, 1985.

Leckenby, Charles Harmon. *The Tread of Pioneers.* Steamboat Springs, CO: Steamboat Pilot, 1971 1945.

Leckenby, Maurice W. *Put a Head on It.* Denver: Sage Books, 1960.

Merrill, Marcellus Samuel. Contributor David Merrill Primus. *Steamboat Springs: Memories of a Young Colorado Pioneer.* Lake City, CO: Western Reflections Pub. Co., 2008.

Perry, Ruth Brown "Ditty." *Marjorie Perry: An Outstanding Woman of the West.* CO: Ruth Brown Perry, 2012.

Pritchett, Lulita Crawford. *The Cabin at Medicine Springs*. New York: F. Watts, 1958.

———. *Crawford Pioneer Tales*. Steamboat Springs, CO: Tread of Pioneers Museum, 2005.

———. *John Daniel Crawford in the Great Good Old Days*. Steamboat Springs, CO: Tread of Pioneers Historical Commission, 1984.

———. *Remember the Old Yampa Valley*. Steamboat Springs, CO: Tread of Pioneers Historical Commission, 1983.

Rockwell, Wilson. *The Utes: A Forgotten People*. Ouray, CO: Western Reflections, 1998.

Summit County Journal. "The Future Metropolis of Routt County, Colorado." 1888.

Valentine, Rebecca. Contributor Travis Thompson. *Beyond the Land of Gold: The Life and Times of Perry A. Burgess*. Acworth, GA: Thompson Media, 2010.

Index

About the Author

Dagny McKinley has an MFA from Naropa University's Jack Kerouac School of Disembodied Poets and Writers and a BA in English from the University of Western Ontario. She studied journalism and writing at UCLA.

She currently works as a freelance writer in Steamboat Springs, Colorado, with a focus on search engine optimization. In an effort to combine her love of literature and poetry with positive change that is being made regarding the environment, she created UndiscoveredEarth.com.

As a journalist, McKinley has contributed to *Modern Dog*, *Ocean Magazine*, *Steamboat Pilot & Today*, the *Denver Post* and the *Aspen Times.* She has also been a contributor to *Mountain House & Home*, *Steamboat Magazine*, *Homelink* and *DCD Home.*

McKinley's books include *Wild Hearts: Dog Sledding the Rockies*, a coffee table book of photography and text based on her experiences working for a dog sledding company; *Lessons My Mother Taught Me: The Good, the Bad and the Questionable*; and *The Adventures of a Girl & Her Dog* (a children's book).

McKinley is the author of "Enna Rose," a long-running column about a free-spirited girl finding her place in a town while her heart lives in nature. Her essay "How Are You?" was recently published in the book *Now: Embracing the Present Moment*. Her short stories have appeared in *Nylon Screams* and *cliteraturejournal.com*. "A Soldier's Peasant" was published in *Di Algo Para Romper Este Silencio*, an anthology celebrating Raymond Carver.

Her play *The Eccentricities of a Butterfly* was produced by the Great American Laughing Stock Company. The story "Pray" was performed by Stories on

Stage in conjunction with the Denver Art Museum's HotDAM events and Denver Center for the Performing Arts actors.

McKinley lives with her dog in Steamboat Springs, where they explore as far as their feet can carry them.